·THE·
DARK SIDE
·OF THE·
EARTH

ROBERT MUIR WOOD

Portraits drawn by Margaret Woodhouse

London
GEORGE ALLEN & UNWIN
Boston   Sydney

**George Allen & Unwin (Publishers) Ltd,
40 Museum Street, London WC1A 1LU, UK**

George Allen & Unwin (Publishers) Ltd,
Park Lane, Hemel Hempstead, Herts HP2 4TE, UK

Allen & Unwin Inc.,
Fifty Cross Street, Winchester, Mass. 01890, USA

George Allen & Unwin Australia Pty Ltd,
8 Napier Street, North Sydney, NSW 2060, Australia

First published in 1985

British Library Cataloguing in Publication Data

Wood, Robert Muir
 The dark side of the earth: the battle for the earth
 sciences, 1800-1980.
1. Earth sciences——History
I. Title
550'.09'03   QE11
ISBN 0-04-550033-9

Library of Congress Cataloging in Publication Data

Wood, Robert Muir
 The dark side of the earth.
Bibliography: p.
Includes index.
1. Earth sciences——History. I. Title.
QE11.W66 1985   550'.9   84-24360
ISBN 0-04-550033-9 (alk. paper)

Set in 10 on 11 point Palatino by Bedford Typesetters Ltd
and printed in Great Britain by
The Anchor Press Ltd, Tiptree, Essex

Dedicated to my parents

Preface

It took ten years before the awareness that something was wrong reached a resolution. On arrival at Cambridge University in 1970 there was talk of a revolution in the Earth Sciences but many of the courses I was being taught had remained unchanged for more than a generation. In the USA it would probably be hard to see what had happened, because history is so readily swept away, but in Britain the massive cultural inertia left a chaos of old and new worlds co-existing. Having begun to unravel this diverse and paradoxical collection of incompatibles I am now grateful that Cambridge University took ten years to wake up to the revolution, and that three separate departments, all studying the same Earth but with different techniques (physics, chemistry and geological description) resisted amalgamation. While living through that period of suspended animation, the period after the cartoon character has walked over a cliff but before he begins to fall, I was privileged to observe the inner workings of a great scientific upheaval.

There was, however, no guide book to what was going on: a moment of enlightenment came in 1979 after a meeting given at the Geological Society of London in which the theory of the expanding Earth was enthusiastically discussed as though the clock had been wound back to 1955. At first the significance of that meeting was a mystery, until eventually I realised that the character who had lost contact with the ground was in fact the distinguished 19th century science of Geology.

There are many powerful parallels between political and scientific revolutions, and two of them are the post-revolutionary desire to rewrite history books to emphasise only the successful struggle and that while the revolution may seem to be concerned with ideas it has far broader implications of people and power. The early 1970s were, in all senses of the word, a 'critical' period. Some of the themes within this book have been developed through testing my critique against several dozen books reviewed for *New Scientist* magazine. A few sections of this book also appeared in an earlier form in articles within that magazine, and I am grateful to Christine Sutton and Michael

v

Kenward for the opportunity both to write these pieces and to incorporate them here. There are several books and articles written on the history of Geology (see Bibliography) that have provided important sources of information although, as those writers have perhaps unwittingly discovered, following the internal history of Geology into the 20th century leads to an inevitable dead-end. There are also a number (perhaps already too many) of good histories of plate tectonics that have saved my research time in reporting interviews with some of the most significant protagonists. However, there is no book that attempts to tell one complete story of the study of the Earth, geologists, geophysicists, dreamers and all, or that explores the cultural significance of the 1960s revolution. My most optimistic hope is that *The dark side of the Earth* will serve to encourage a debate as much about the past as about the future shape and constitution of the Earth Sciences.

There have been many inspirational conversations: the idea for the book came from a lunch-time chat with historian Jonathan Steinberg at Trinity Hall, Cambridge. A whole new avenue was suggested by another historian, Arnold Harvey. A programme for BBC Radio 4 in summer 1984, organised by Martin Goldman, provided me with the opportunity to quiz a number of important figures including Sir Kingsley Dunham at Durham, Keith Runcorn at Newcastle, Sir Harold Jeffreys and Brian Harland at Cambridge and John Tuzo Wilson at Toronto. A period of reading and working in the history of science was afforded by the Master and fellows of Trinity Hall through a research fellowship from 1977 to 1980. From 1981 it has been possible to write, thanks to an Alphatronic microcomputer, amidst the innovative consultancy projects undertaken for Principia Mechanica Ltd. I praise the tireless librarians at the Cambridge University Library and at the library of the old Geology Department and I bless the inventor of the photocopying machine. Russian translation was kindly carried out by Patrick Miles and Jean Agrell. I also thank Roger Jones of George Allen and Unwin for giving me a contract to write a completely different book and not being too upset when this present one arrived.

Susan Littley helped in more ways than she realises. There are, as a reader will find, no women in this history (and that is in itself a story), and yet my own interest in Earth studies comes from my great aunt Helen Muir Wood, a superb palaeontologist, supporter of Wegener's drift, and administrator at the British Museum for Natural History in South Kensington, who slyly set up exhibitions showing moving continents in the 1950s. A number of friends have read and commented on the book at various stages including: Geoff King, Gordon Woo, Willy Aspinall and Helen Wood. The criticisms made of an earlier draft by readers of the book, Bob Hazen and Peter J. Smith were extremely apposite. Any errors or too simplistic elisions of complex debates are of course my own.

Robert Muir Wood
4 July 1984

Contents

Contents

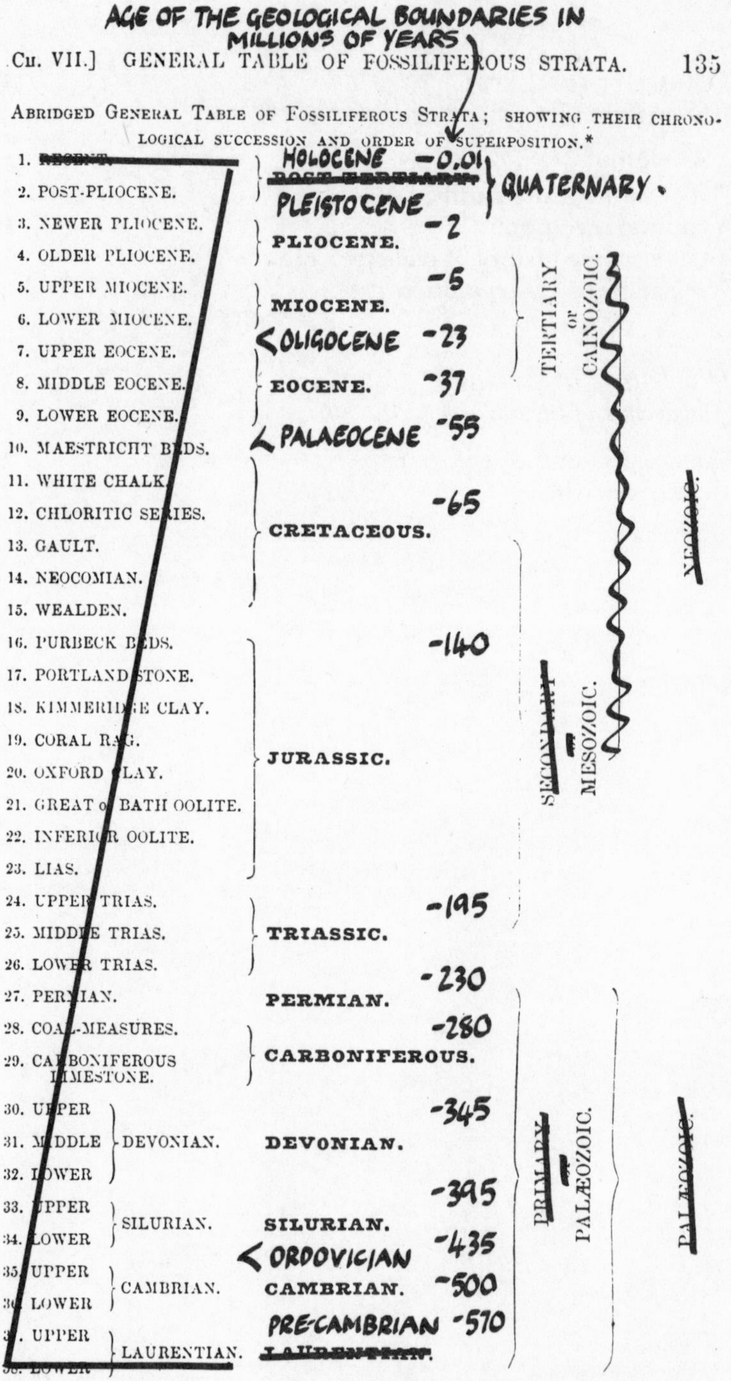

AGE OF THE GEOLOGICAL BOUNDARIES IN MILLIONS OF YEARS

ABRIDGED GENERAL TABLE OF FOSSILIFEROUS STRATA; SHOWING THEIR CHRONOLOGICAL SUCCESSION AND ORDER OF SUPERPOSITION.*

Strata		Period	Age
1. ~~RECENT.~~		HOLOCENE	-0.01
2. POST-PLIOCENE.		~~POST-TERTIARY.~~ PLEISTOCENE } QUATERNARY.	
3. NEWER PLIOCENE.		PLIOCENE.	-2
4. OLDER PLIOCENE.			
5. UPPER MIOCENE.		MIOCENE.	-5
6. LOWER MIOCENE.			
7. UPPER EOCENE.		< OLIGOCENE	-23
8. MIDDLE EOCENE.		EOCENE.	-37
9. LOWER EOCENE.			
10. MAESTRICHT BEDS.		< PALAEOCENE	-55
11. WHITE CHALK.		CRETACEOUS.	-65
12. CHLORITIC SERIES.			
13. GAULT.			
14. NEOCOMIAN.			
15. WEALDEN.			
16. PURBECK BEDS.		JURASSIC.	-140
17. PORTLAND STONE.			
18. KIMMERIDGE CLAY.			
19. CORAL RAG.			
20. OXFORD CLAY.			
21. GREAT or BATH OOLITE.			
22. INFERIOR OOLITE.			
23. LIAS.			
24. UPPER TRIAS.		TRIASSIC.	-195
25. MIDDLE TRIAS.			
26. LOWER TRIAS.			
27. PERMIAN.		PERMIAN.	-230
28. COAL-MEASURES.		CARBONIFEROUS.	-280
29. CARBONIFEROUS LIMESTONE.			
30. UPPER	DEVONIAN.	DEVONIAN.	-345
31. MIDDLE			
32. LOWER			
33. UPPER	SILURIAN.	SILURIAN.	-395
34. LOWER		< ORDOVICIAN	-435
35. UPPER	CAMBRIAN.	CAMBRIAN.	-500
36. LOWER			
37. UPPER	LAURENTIAN.	PRE-CAMBRIAN ~~LAURENTIAN.~~	-570
38. LOWER			

TERTIARY or CAINOZOIC. — ~~SECONDARY~~ = MESOZOIC. — ~~PRIMARY~~ = PALÆOZOIC. — ~~NEOZOIC.~~ — ~~PALÆOZOIC.~~

* For a more detailed and extended list see Elements of Geology, 6th edit. p. 102; and Student's Elements, p. 109.

Frontispiece From *Principles of geology*, 11th edn (Charles Lyell 1871) with 20th-century refinements

Introduction

Geology is too defective! We hardly know more than a few parts of Europe. As for the rest, including the sea-bed we shall never know about it.   (Bouvard to Pecuchet – Flaubert 1881).

For someone who had never studied the subject, even scientific literacy would be unlikely to provide the name of one important 20th-century geologist. It was not always this way. For a period during the early 19th century, geologists were the most famous of all scientists and Geology the most fundamental science, involved in the reconstruction of man's concept of time, telescoping out from the measure of human ancestry towards eternity. It was the Association of American Geologists which grew to become the American Association for the Advancement of Science. Great geologists such as William Buckland and Charles Lyell were the primary inspiration equally of Charles Darwin as of Alfred Tennyson.

The heroic age passed on. Geology became a forgotten and remote territory, a country that history had bypassed. There followed almost a century of silence. Suddenly at the culmination of the 1960s there was an eruption of activity. In an age of affluence and certainty, to a background of rock-music and space-travel, there was a new plastic art of global geography, and of 'plate tectonics'. A young generation of marine geophysicists, lean and tanned from long voyages exploring the oceans, looking more like surfers than scientists, talked of India moving north and America gliding west as though they were ships passing in the night. The moving continents captured the popular imagination and soon became more familiar to schoolchildren than the rocks themselves.

Had Geology at last made a come-back? That was how it seemed and yet a simple dip into the history of this self-proclaimed scientific revolution reveals many problems. The greatest prophet of continental drift, Alfred Wegener, was a meteorologist; continental drift first reached England in 1921, yet it was 14 years before the Geological Society of London arranged a debate on drift. In 1949 in America, drift

1

was called a 'leftist' theory; in the 1950s it was claimed that a US Geology lecturer could be dismissed for teaching Wegener. The new theory of plate tectonics was compiled by geophysicists with whom geologists had sustained a considerable antipathy. Even in 1982 the majority of Soviet geologists still did not support plate tectonics.

This book is not a standard history of science, for the story passes straight into the present day. It also extends out to the margins of science, into the borderlands of pseudo-science, where Wegener's continental drift theory had for many decades reposed. To encompass the creation of the noble idea of the Earth, an idea that evolved as much through the bold imagination as through scientific investiga-tion, requires history painted with large strokes on a broad canvas: reaching back in time to the birth of Geology and forward to the 1980s; extending across many countries: America, Britain, France, Germany, Holland, South Africa, Australia, Canada and Russia. The theatre of global theories was forever on the move: from the Alps to the jungles of the Dutch East Indies, from high on the Hawaiian volcanoes out into the Red Sea. The history follows those outsiders who challenged established science, those leaders who redirected the course of research and the ever-present conflicts that were fought within and between the sciences.

For the discovery of the Earth and the creation of the Earth Sciences were consistently resisted by the geological establishment. Following a bitter struggle with the physicists over the age of the Earth, Geology failed to adapt to the new style of geophysical investigation. The decline of a science can be irresistible. Like a dynasty without the ambition of the first monarch the creative force became replaced by ritual and repetition. An alien science had been developing on the margins of Geology for almost a century. A generation after the Second World War this new scientific structure challenged and overwhelmed the old.

1

The new Stone Age

The birth of Geology

The tale I wish to tell concerns the conquering of the Earth: a victory more total than any gained with military might, yet the tale is a humble one – a history of those who subdued the planet through curiosity. Such a theme, the dominion over the Earth, has powerful resonances with mythology. The story, too, has ingredients more familiar in sagas of the gods: the ruler of the world, a primal tribal enmity, the marriage between opposites, the over-reaching ambition, the fatal flaw. The history begins at a time when a formative battle for power was about to reach a resolution. The turning point came around 1800.

In the 18th century investigations of the Earth were undertaken without a central institution, without an organised framework of studies, in a mix of natural philosophy and observation, divine explanation and speculation. There was only one science of the earth and that was Mineralogy, taught in mining schools in France and the German-speaking states but largely ignored in England and Scotland. During the late 18th century, at the Mining School at Freiberg, Mineralogy made a vainglorious bid to rule the Earth. Professor Abraham Gottlob Werner produced a powerful alliance between Mineralogy and the Book of Genesis that explained the origins of the Earth according to a history later termed 'neptunism'. All of the rock formations had crystallised out of a primeval briny bath, that as it had gurgled into subterranean cavities had also carved deep valleys and lain down vast tracts of sediments. All the incidents within the Creation of the Earth were based on simple mineralogical principles. The outermost stratal rind, of crystalline rocks at the base overlain by a variety of layered secondary and friable tertiary formations, was a product of the same 'order of crystallisation' that could be observed in the evaporation of sea water. The Flood was Creation's encore; volcanoes the local result of subterranean coal fires. The empire of earth studies, ruled by mineralogists, in which fossils

3

were as mere idle gossip in the Story of Creation, was to be known as 'geognosy'.

The most coherent opposition to Werner became centred around the convoluted writings of the Edinburgh philosopher and industrialist, James Hutton, whose followers became known as 'vulcanists'. Hutton and his close friends David Hume and Adam Smith as part of the Scottish Enlightenment sought to find in the operations of the natural world, cycles of activity and equilibria of opposing forces. Their religious faith was deist – the proof of the ultimate wisdom of God was that he had created an entirely self-sufficient Natural World. It was an imperfect clockmaker who had to regulate the hour of his timepieces. Hutton's close observations of the rocks were used to support his farmer's world-view. He wrote a treatise that was to make 'Philosophers of husbandmen and husbandmen of philosophers'. Rocks must continually decay to provide the new soil to support plant life. The destruction of rocks is therefore just one part, the visible half of a cycle that must also involve rock creation. Decay through the action of water is matched by reconstruction through the action of heat, on the sea floor and deep into the earth.

Neptunism and vulcanism were both theories of epic scale, and yet their names give a false impression of direct confrontation – as the power of water against the fury of fire. The central tenet of neptunism was that the *Creation* of the Earth was directly recorded in the configuration of the rocks; of vulcanism that the configuration of the rocks provided insight only into the operation of the *process* of the Earth. Thus, viewed as the focus of wide-ranging opinion, an inclination towards neptunism was encouraged by studying mineralogy and reading Genesis; a tilt towards vulcanism by observing, and reading with an innocent heart, the landscape of rocks. Vulcanism was a gentler, more mystical Eastern cosmology in opposition to the apocalyptic biblical neptunism. In the war of 'how it ran' versus 'how it began' the forces concentrated debate on a material rather than theoretical level (around 1810) in the field of basalt. The arguments as to the origin of this dull, black, featureless rock finally undermined Neptunism. The demonstration that the basalt of Antrim was identical to the material that had flowed in prehistoric times out from the volcanic craters of the Massif Central and Eifel vindicated vulcanism. To future geologists this contest stood at the Creation of their Science; Hutton's victory was celebrated, neptunism denigrated. But cosmologies with the mythic power of Werner's do not simply fade away. In Sweden and the German states geognosy and the Wernerians lingered on opportunistically absorbing the new discoveries of Geology. In England the philosophy of neptunism was to reappear in an adapted form with a new name; yet not before the great Geology circus had got underway.

Away from the wild polemics of the Edinburgh vulcanists and neptunists (for the city was a hotbed of Wernerians), and with no great mineralogical tradition to impose its method, London provided the clean slate upon which the constitution of Geology could be chalked. The word 'geology' had been adrift for the previous 150 years, and had been sparingly used by Hutton, and thus the founders of the 'Geological Society of London' in 1807, although preferring utilitarian goals rather than cosmological controversy, tilted towards the vulcanists. (That Hutton is the true 'founder of geology' is now ineradicable: in 1947, in true geological fashion, this message was chiselled onto his tombstone.) The society was founded, as were all societies at this dawn of the great age of institutions, with a few gentlemen meeting over dinner (at the Freemason's Tavern in Great Queen Street). Between mouthfuls of food the word 'geology' was tested and found to be ripe. A new word performs many mysteries in dissolving the past and evoking the future. The 'rite of passage' of Geology, of which the founding of the Geological Society of London was just one ceremonial act, took place during this, the first decade of the new century. There were three phases to the rite; all marked by an indulgent rhetoric.

The reaction that served to condemn the past: previous earth investigations were marked by 'A species of mental derangement, in which the patient raved continuously of comets, deluges, volcanoes and earthquakes.'[1] The fever was now subsiding, the good and sane Doctor Geology had arrived.

The exchange of credentials with the rival sciences that could serve to diminish and debase them: 'The connection of mineralogy to geology is somewhat of the nature of that of the nurse with the healthy child, born to rank and fortune'.[2] While Mineralogy could be disparaged for its lowliness and servility and its treacherous involvement with the neptunists, cosmogony (the study of the origins of the Earth and the universe) was the chief focus of scorn and derision. Cosmogony! Sprawling and fantastic. Unfashionable theories, such as neptunism, were dismissed as 'speculative cosmogonies'.[3]

The endeavour: Geology was to be the art of the possible. Theory was cosmogony. Theory was to be outlawed, at least at the beginning. 'The fabulous and romantic age of geology may be said to have passed away; its disciples, no longer engaged in the support of whimsical theories direct all their attention to the discovery of facts, and to their application to purposes of extensive utility'.[4] The infant Geology was to be reared on hard-won facts rather than seduced with speculation. Geology restricted itself to 'the different mineral substances that compose the crust of the earth'[5] and performed the study through examining 'the relative position and mode of formation' and the 'relative structure and ages' of the rocks. Most important to geology's constitution was the recognition that it was the study only of the accessible; of the surface of the Earth. Unless all theories could be tied to the rocks as experienced by the geologist they were just speculations – drifting free – cosmogonies.

At the end of this endeavour there was to be a final goal: 'the theory of the earth'. Playfair, speaking for all geologists, affirmed the credo that: 'if the face of the earth were divided up into districts, and accurately described we have no doubt that, from the comparison of these descriptions, the true theory of the earth would spontaneously emerge without any effort of genius or invention. It would appear as an incontrovertible principle.'[1] This was a crucial statement of ambition and intent. It was to lie at the very core of geological philosophy. The house of Geology was to be built stone by stone; its foundations were to be wide and imbedded in the solid rock. Without any architect, the house – the theory of the earth – would be finished when all the stones had been laid on top of one another. This was the plan.

Unknown to the founders of the Geological Society of London, their pragmatism had already been vindicated by the discoveries of one William Smith, drainage engineer. Between 1791 and 1799, as Smith supervised canal excavations across the length and breadth of England, he noticed that the fossils found within a particular stratum provided a unique marker that could be followed along the walls of the cut and, even after the layer had disappeared, could be rediscovered at some distant outcrop. The key to this remarkable insight came from the nature of the enterprise. Fossil stratigraphy is a direct product of canal building. The excavation provides a slow-moving traverse; the rocks displayed are superficial and of very limited extent, and for much of England are of soft sedimentary material not otherwise revealed at the surface. The canal engineer must make a study of the rocks because the very existence of the canal is dependent on the permeability of the underlying strata or the availability of impervious material nearby. Only when Smith produced his detailed geological map of England in 1815 did his method become known and, subsequently in the late 1820s, lauded as providing the long-awaited key to the historical mysteries of the rocks. As William Smith was more concerned with puddling clay than with earth history he was never invited to join the clean-booted gentlemen fellows of the Geological Society.

After William Smith the 'accurate description of the earth's districts' was a mapping programme. If Playfair's endpoint was ever in view – that a theory of the earth would spontaneously present itself at the completion of such work – then it was soon forgotten as the enterprise became ritualised as the 'work of the geologist'. In this form it has survived remarkably unaltered. Thus Geology as it was practised became less concerned with problems to be solved, but instead, as in canal building, with certain traditional procedures for documenting the distribution of rocks at the surface.

The psychology of Geology

At its origin Geology took great pride in the organised objectivity that it offered to the investigation of the earth. This objectivity arose, not because man had now successfully conquered the earth, but on the contrary, because a method had been found by which to limit and order the investigations, thereby to tackle only those problems that were visible and soluble. It was the very sobriety, the economy of the attack that allowed the puritan Geology to be successful and to lampoon so pitilessly the nightmarish indulgencies of the cosmogonists. Yet this faith in objectivity served to obscure a profound set of problems. For no other object in the universe dominates human perception to the extent of the Earth. This dominion is all the more powerful because it is unperceived; the Earth provides the fabric on which all experience is located. To sever even for an instant the implicit bond with the Earth can be traumatic. Charles Darwin, caught in the great Chilean earthquake of 1835 wrote 'A bad earthquake at once destroys our oldest associations . . . one second of time has created in the mind a strange idea of insecurity which hours of reflection would not have produced.'[6] To be reminded of the Earth requires that its solidity and permanence is taken away. To perceive the Earth as an object and not as the all-pervasive subject is in effect to be weightless.

The bond with the Earth was at the very heart of the science of Geology and yet was never identified as being implicated in that science's structure. As a result of this failure to separate man's experience from the object he wished to study, Geology became articulated and disposed in the measure of man more than any of the other natural sciences. To understand Geology, its history, its position within the wider culture, its hesitations and successes it is necessary first to leave behind the objects of its enquiry: the rocks, minerals and fossils; the airy crags, cliffs and quarries, and to turn instead to a study of its most important structural constraint: the role and the location of man the observer.

To propose that Geology cannot be understood without psychology, is a long-forgotten priority but not an original one, for Charles Lyell, in the archetypal geology text *Principles of Geology* (1830), dedicates an enlightened introductory chapter to dispelling 'Prejudices which have retarded the progress of geology'. These prejudices have all arisen from the location and measure of man. They include the desire to accommodate all past times to the scale provided by human memory, the experience of geological processes being always land-based and including nothing of the underwater (Lyell believed that 'an amphibious being, who should possess our faculties' would be better equipped to be a geologist), and always surface-based and including nothing of the deep underground.

Lyell's purpose in listing these prejudices is not to enquire too deeply into the distortions imposed upon the construction of a Science of Geology. He presents them, like a priest summoning up bad spirits, in order that they might then go away. To join battle with the 'prejudice of human-time' was to prove the great crusade of the heroic age of Geology. Geologists were engaged in the struggles to provide a

Charles Lyell, 1797–1875
British geologist and philosopher

history of the world to rival the written or oral tales of the ancestors. They were to produce a History so powerful and so mythic that it was to replace the story recorded in the Book of Genesis, and forever to discredit fundamental religion. Lyell was to bless Hutton's uniformitarianism by which Geology could offer itself scientific access to the past. Activity had always continued along those lines observed in the present. As in the battle of neptunism and vulcanism, human memory contained in ancient writings was no longer considered sufficient to provide the beginnings of the world; in its place human experience and knowledge of the earth was to provide a cipher by which Earth's history was decoded. The three great philosophical questions of origin, the origin of the world, the origin of life, the origin of man, were all to be wrested from the Bible and placed in the ambit of Geology.

The triumph of Geology in rewriting the history of the world proved that the prejudice of time could be defeated. Yet reason itself was incapable of engaging with those other prejudices formed by the inaccessibility of the underwater or underground worlds. For as long as geologists were content to limit their enquiries to unravelling history and mapping the local rock formations, inhibited from wider contemplations by the early tirades against cosmogonic theory builders, these problems of inaccessibility would never be challenged. Yet the motives for the caution expressed at the origins of Geology were to become forgotten. For after Lyell, the overall structure of the science became established more or less according to his design. Therefore after Lyell the 'prejudices of geology' suffered no more investigation. The patient had passed through the crises of childhood;

ever after these distortions would be buried within the unconsciousness of the adult science.

Beyond Lyell's unresolved shortlist there was one other greater prejudice that remained unperceived, yet that had already determined much of the stratagem of Geology. It was the prejudice of location and scale. The Earth was unknowable, not just because parts of it could not be reached but, more importantly, because it could not be realised. All sciences begin with a phase of visual exploration and identification. At the limitations of the naked eye the lens takes over: the magnifying glass, the telescope, the optical and electron microscopes all perform identical functions in enlarging the world beyond the physiological limitations of the eye. Yet the geologist is crammed onto the object he wishes to study. The optical instrument he requires is one that can correct the aberration of his absurd position squat up against the Earth. In the absence of such an ultimate fisheye lens, he extends his investigations with a hammer. The hammer was to become the emblem of the geologist: 'Jones was a geologist in the time-honoured tradition whereby most of the work is done on the actual outcrops. Referring to a particular section he would claim that he had hammered every inch of it, and that if anyone wanted to refute his conclusions he would have to do the same'.[7] Such a quotation would be hard to date; the geologist's culture survived unscathed for 150 years (the description is taken from an obituary written in 1982). Yet the hammer-blows the geologist has always directed at the earth accompany a deeper frustration. For at the base of Geology there is this extraordinary and profound problem: that the Earth has no handles. Plants, animals, even stars, are all discrete objects; their study can get underway without any agonising over what is to be studied. Absolute size is of no concern; it is the relative size to the observer; the Moon is as a sea-anemone, the stars as corpuscles viewed down some microscope. In Geology there was only one discrete object, planet Earth: an object that Geology at its foundation denounced as being beyond comprehension.

As the hammer made contact with the rock, the continuity of the Earth's surface could be forgotten; the most important epistemological problem in Geology, that of determining how the study should get underway, was met not with calm contemplation but with the smashing of the hammer. For once the smashing was completed, the problem vanished. Now in this rock, this sample, there is a discrete entity to be studied. The hammer itself has been borrowed from the blacksmith or the stonemason – who are creating objects for the use of man. The borrowed tool can only create according to its function: the size of the horseshoe or the stone-slate. The rock fragment became the object under study as if it were made predeterminedly discrete when it was simply a sculptured creation. Of available sizes from the dust grain to the continent, the geologist recreated the Earth according to the size of his own fist. The residual memory was at work. Stone Age man was reborn, busy manufacturing artefacts, at the beginning of the 19th century.

A tale of two Earths

The early self-definition of Geology proscribed extending the scope of the enquiry beyond that which could be realised by a man. This scope is the extent to which an investigator could be said to have knowledge of the rocks of a landscape. The scope is visual – a view of encircling ranges, from a hilltop or a mountain peak. The map could be extended across a county, state or country, but still the scale, the scale of visual understanding, remained the same. The existence of this measure, the scale of perception, can be found from investigating within the 19th century the largest feature to which geologists were prepared to give their own independent name and for which they sought an explanation: the great mystery of the Age of Romanticism, the 'mountain range'. The mountain range was to become the upper boundary of abstraction.

In 1859 James Hall, State Geologist of New York, found, buckled up in the Appalachian Mountains, there were 40000 ft of shallow marine sediments, while following these layers to the west, beyond the mountains, the unfolded rocks of the same age were only a few thousand feet thick.[8] Either the mountain range was predestined or, as Hall explained it, the big thickness of sediments pulled down the crust to form the mountains. James Dwight Dana, American mineralogist, geologist and philosopher, in 1873 sandbagged this explanation calling it a 'theory of mountains with the mountains left out', and in blessing some form of predestination termed the thick pile of sediments a 'geosyncline'.[9] Thereafter on both sides of the Atlantic the problem of geosynclines was to remain the fundamental problem at the uppermost scale of Geology; the origin of these features became the impasse on the journey towards a theory of the earth.

Those geologists who stretched the scope of Geology, who pushed against the walls of the bubble of perception, were those who, in place of detailed mapping and sample collecting, preferred to travel. Grove Karl Gilbert, while trekking through the Western States in the 1870s, came armed with a surveyor's plane-table and saw within the barren uplands of the Henry Mountains stretching before him, that a giant volcanic piston had raised the whole massif through infilling a vast internal balloon of molten rock.[10] To pass beyond the scope offered by the American West, of deep raw cuts in the outer skin of the Earth, requires leaving the planet's surface behind. Eduard Suess, the greatest geological syncretist of the late 19th century, began his comprehensive four volume overview of Geology *The face of the Earth* with a vision of the Earth first glimpsed by a traveller arriving from outer space. Alfred Wegener, the great promoter of continental drift during the early 20th century, gained inspiration as a young man from flying in a balloon high above the Hanoverian countryside. The South African, Alexander Du Toit, almost unique among geologists in the 1930s as an advocate for continental drift, 'delighted in air-travel, for it gave him a broader view of earth phenomena than he could get with his feet on the ground'.[11]

For the yet more elevated view one has only the bubble-gum quotations of the astronauts. As the geologist Harrison Schmitt voyaged aboard Apollo 17 he looked back at the Earth and reported to Mission Control, 'I didn't grow up with the idea of drifting continents . . . but I tell you when you look at the way the pieces seem to fit together, you could almost make a believer out of anybody'.[12] All that can be seen from the surface are the elevations of the topography – and hence the geological theories that emerged from these views stressed the vertical. The Earth of horizontal forces has been confirmed by images of the 20th century, from canal builders to space travellers, the views are now from above.

This subjectivity of scope is fundamental: *Homo sapiens* is chained by gravity as a prisoner on the planet's surface. Nowhere are the consequent prejudices clearer than in the original establishment of the relationship between Geology and Astronomy. In Chapter One of the *Principles of Geology* Lyell writes 'Geology is intimately related to almost all the physical sciences as history is to the moral'. The significance of the qualification is revealed with the list: 'it is desirable that a geologist should be well versed in chemistry, natural philosophy, mineralogy, zoology, comparative anatomy, botany'. For a geologist in the image of Lyell the relationship between Astronomy and Geology was simple: the sciences had no contribution to make to one another. This was for a variety of reasons. The geologist roamed across the landscape by day, his eyes searching the ground, while the astronomer hid from the daylight, remaining at home to stare deep into the night sky. Culturally, behaviourally, a geologist was more likely to have parallel interests in natural history; an astronomer in mathematics. Occasionally some report of a meteoric stone threatened to implicate careless geological activity in the movements of the heavens; but their absence from Lyell was proof enough that meteorites were not Geology.

More important than the cultural disparities, the retrenchment of the 18th-century studies of the Earth to form Geology purposefully drew away from Astronomy because the observations of the heavens had proved a major source of speculation and theory. Stars were mysterious, evanescent, the stuff of dreams and astrology. Lyell could chide astronomers in his conclusion to *The elements of Geology* (1838) in saying that 'Astronomy had been unable to establish the plurality of habitable worlds throughout space, however favourite a subject of conjecture and speculation; but geology . . . has demonstrated the truth of conclusions scarcely less wonderful – the existence on our planet of so many habitable surfaces, or worlds as they have been called, each distinct in time . . .'. The despised cosmogonist, like the astronomer, sat at home, his fantasies indulged by moonshine.

If one was to attempt to find the boundary between the two sciences then it seemed to those involved to be crudely Aristotelian – the Earth demanded a different science from that of the heavens. Yet as Geology was the science of the local earth according to man – the egogeocentric Earth – the boundary between the sciences in practice was where the atronomer's telescope, coming to the edge of the dome of the sky, grazed against the Earth's convexity. The seam between Geology and Astronomy was the visual horizon. To geologists the difference between their own territory and its neighbour was that between matter and spirit; the boundary was scarcely remarked and never crossed. Only to those whose imagination could leave behind the confines of the visual landscape was there no such seam, but instead a yawning chasm, containing the Earth that could not be seen – that plunged beneath the horizon, that extended over whole continents and oceans – a gap so wide that it reduced the concerns of Geology to the merely parochial. This unseen Earth required imagination to perceive and it was exactly the illusory imagination that was damned as cosmogony, neoplatonism, metaphysical. One of the earliest books that had taken as its title *The theory of the Earth*, written by Robert Burnet in 1684, had revelled in the power of the imagination to conquer the Earth: 'there is no Chase so pleasant, methinks, as to drive a Thought by good conduct from one end of the World to the other'. Burnet provided the most powerful invocation of a global scope to the study of the Earth, that pre-existed Geology. His book was reprinted many times through the 18th century and was much admired by both Coleridge and Wordsworth. The final publication date, of 1828, marks the point at which the new orthodoxy of Geology had become fully established.

The intellectual territory between Astronomy and Geology, along with other inaccessible marches and no man's lands, is without name, for to name it would have been to legitimise the enquiry. It remained largely unrecognised through the illusory nature of perception and the chameleon properties of the word 'earth'. There are many 'earths' (17 in the *Oxford English Dictionary*), but essentially only two. The little 'earth', the lower case earth, is the earth of our culture and perceptions. It is the soil, the ground, the earth that a soldier is sent to defend, for his ancestors' bones, which have rejoined the earth, thereby uniting it with its inhabitants. This primitive earth, the earth that is the foundation of all settled peoples, is mother and home: the source of wisdom, all that is substantial in the face of illusory goals. Then there is the second, ineffable Earth; the Earth that is the counterpart to Heaven. Such abstract Earths receive little physical confirmation. The table-top globe may guide the projection but provides a poor image for the vastness of a planet.

The British astronomer Fred Hoyle wrote for a BBC radio broadcast in 1950 'Once a photograph of the Earth, taken from outside is available a new idea, as powerful as any in history will be let loose'.[13] The photographs of the retreating Earth, taken on the Apollo moon-flights (silky and blue, the bulk of Africa painted in the hot colours of scorched soil, framed on a million bedroom walls), have provided the first substantial image, an image of majesty, and vulnerability. It is this new image of the Earth that is the pillow-mate; a fragile planet to be loved and cherished.

In the early 19th century, as geology chose to stimulate investigation of the little earth, so it could claim to be 'the science of the earth'. The gap between Geology and Astronomy that was the study of the big Earth, wâs a linguistic chimera; a geologist could readily argue that it did not exist. The potentiality of this idea of 'Earth' was only revealed in the 1950s with the creation of a new research structure termed 'Earth Sciences'.

With the eventual victory in the 1960s of the new scale of perception, the scale of the planet, the first brave dream of Playfair, of arriving at a theory of the Earth through patient mapping, was finally overturned. Lyell had recognised the prejudices of inaccessibility implicit in Geology but had not the wherewithal to overcome them. Other prejudices, the disdain of one scientific discipline for another, along the lines of Lyell's disparagement for Astronomy, were to become of increasing importance. A bitter conflict was to develop into a permanent rift between Geology and Physics. Geology, as the science of the little earth, continued in the manner in which it was founded. Geologists could either concentrate on the smaller-scale surface phenomena which their science had been developed to tackle, through the hand specimen and the field map; or else attempt to extrapolate from their observations into those hidden worlds into which, as Lyell had identified, their science had no access. Such extrapolation was always uncertain. For in the midst of the objectivity claimed by the founding fathers there was a human science, based on observations made almost at ground level, on rocks that fitted into the human palm, all within a domestic science: the study of man's home territory. If man had just arrived from outer space then it would be obvious that the primary scale of study was that of the Earth as a planet. Until man could return from outer space, that image of the Earth and that scale of the planet could only be reached through that chrysalis of myth: the imagination.

2

The big apple

The web of Geometry

It is a warm August afternoon. The car-tyres are racing over the cobbles; the sound is of potatoes tumbling down a chute. In the midst of the traffic whirlpool a stroll around the great stone vault of the Arc de Triomphe reveals a recurring view that slots into place 12 times as the converging parallels of yet another road come into view. From the top of the arch where the traffic sound is a buzz and the city skyline blurs into the haze, the realisation dawns that you, the observer, are standing at the focal point. The Arc de Triomphe is elevated by the profusion of roads that lead to its base – a superiority of position as if it were constructed on the summit of a hill. And yet the view itself is not the intended effect, nor the multitude of views of the arch itself, seen on approach, crowning the end of some radial highway. Place d'Étoile, the 12-pointed star, can only be studied on a map or observed from the sky.

This star-view was first seen on the 1:5000 plans, strapped to screens around the working apartment of Baron Georges Eugene Haussmann, Prefect of the Seine. The star was the culmination of his efforts in the mid-19th century to recreate Paris. It was to be translated into action through his patron, Emperor Napoleon III, and realised like a medallion of honour upon the city's fabric. Starting only with a crossroads, Haussmann added three new connections out of utility and five out of no other desire than that of cartographic symmetry. For Haussmann was a sculptor in the medium of cities; the ultimate cartographer who reconstructs the landscape in order to improve the appearance of the map. His city planning was never without controversy: in 1867 a correspondent in *Le Temps* remarked that 'If it was not a plot directed against the Parisians themselves, one must suppose that Paris planning for fifteen years has been given over to geometricians who have amused themselves drawing lines indiscriminately from one point to another'.

In 1866, as Haussmann was carving geometry into the street map of

14

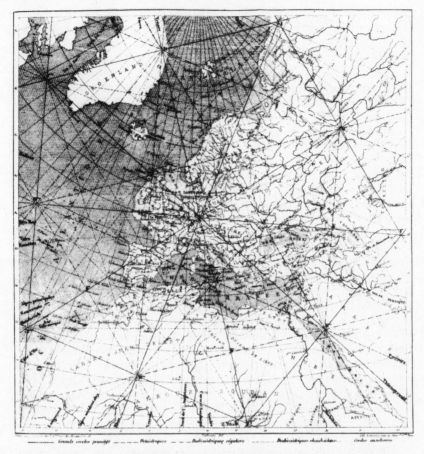

Elie de Beaumont's 1852 map of his réseau pentagonal superimposed on Europe

Paris, Jean-Baptiste-Armand-Louis-Léonce Elie de Beaumont published a geological map of France[1] on which was marked a fragment of the great web of geometry that he believed permeated the Earth. Geometry was implicit in the crust although only faintly revealed; the maps must therefore serve for elucidation.

Elie de Beaumont had attended the College Henri IV in Paris a decade before Haussmann. Outstanding in Mathematics, he passed through the École Polytechnique, the college of engineering excellence, graduating top of his class in 1819. He then made the unusual and formative move into the École Royale des Mines. His timing was flawless. After years of petitioning, the French Government had recognised the connection between the advancement of geological investigations and a nation's industrial performance. In 1822 Elie de Beaumont was invited to travel to England to investigate mining and industry. The Napoleonic wars had interrupted the interchange of ideas and the visit therefore provided an opportunity to bring back the new English science of Geology, and thereby to become a founder of *La Géologie Française*.

In 1823 when Elie de Beaumont made his visit catastrophism was all the rage; the Biblical Flood was no longer a unique re-emergence of the geognosist's primeval ocean but just the latest in a whole series of devastations and inundations that had drawn down the curtain at the end of each geological epoch. The most conclusive evidence for such catastrophes came from the fossils; in 1817 the Parisian anatomist, Georges Cuvier, had been the first to announce the recurrence of such disasters throughout geological history.[2] Cuvier had successfully transformed the single act of Creation in Wernerian geognosy into a multiple catastrophism. Elie de Beaumont never rebelled against his own Parisian baptism as a catastrophist. Yet his immediate concerns were now those of the geological map-maker.

The rolling hills of England were formed from sediments almost all of which contained fossils, and therefore termed in the geognosist's nomenclature as the secondary and tertiary formations. Mapping these strata presented few problems; the map could be continued across the Channel into the identical formations of the Paris Basin. But further to the west in the old rugged landscape of Brittany, to the east in the forests of the Vosges and to the south around the volcanoes of the Massif Central the sediments were replaced by crystalline rocks, deformed rocks, rocks without fossils, rocks of the primary formations. The limits of the English experience defined the territory of the French endeavour: the catastrophes that had ended the geological eras, the unravelling of the primary formations – these were to make up the French contribution to Geology. The key to this opaque history was to come not from palaeontology but from the original and archetypal science of Geometry.

After five summers of meditation, undertaken while roaming his allocated eastern half of France, Elie de Beaumont presented the new theory on 22 June 1829 at the Académie des Sciences.[3] The catastrophes, the revolutions, were associated with escaping vapours and the sudden upheaval of colossal mountain ranges. Since each phase of mountain building occurs as a result of a single mechanical disturbance, he argued borrowing from Humboldt, all the mountains of a common period share the same orientation. Thus the primary rocks could be correlated from one ruined mountain range to another. These were no mere Acts of God; but spasms of corrugation brought about when the forces of collapse became irresistible. Mountains were the proof that the Earth was undergoing an irreversible contraction and decline.

A belief that mountains were the product of a slimming Earth was not new. Newton in 1681 had described the landscape as forming through 'ye breaking out of ye vapours from below before the earth was well hardened, the setting and shrinking of ye whole globe after ye upper regions or surface began to be hard'.[4] Elie de Beaumont had simply punctuated this tendency: in his first lecture he identified six primary orientations indicating six catastrophic phases of mountain ridges bursting through the clouds. By 1833[5] the number had increased to 12 and for much of the rest of Elie de Beaumont's life, continued to grow at the rate of about one per year. Such a profusion of orientations threatened to enmesh him.

Fortunately in the École des Mines where Elie de Beaumont was now a teacher, there was a discipline that could hope to make sense of this errant geometry. Crystallographers found it convenient to map the faces of a crystal onto the surface of a sphere. Suppose that sphere was the Earth. Through crystallography the mountains could be slotted together. Elie de Beaumont first believed that the symmetry was going to be that of the eight-sided octahedron but when the octahedron did not fit, he tried some additional faces, until finally abandoning this form. On 9 September 1850 he presented to the Académie des Sciences his revised geometry: the *'réseau pentagonal'*.[6] The edges of the 20 sides of the icosahedron were continued to form 15 great circles girdling the planet. Where these circles intersect, 120 triangles are formed that account for 'the 24 mountain systems' of Europe. The early-19th-century discovery of geometrical crystallographic laws, implicit within all minerals, demonstrated that the force of crystallisation was a universal force, like the force of gravity. The icosahedron was the Platonic, and therefore perfect, solid closest to that most ideal of all geometrical shapes: the sphere. Thus the combined and opposing forces of gravity and crystallinity would tend towards the icosahedron.

By 1866 he had learnt how to display his crystalline edges onto the geological map of France, that he had first presented, free of the cage of lines, in 1841. In Elie de Beaumont's plan, geometry would ultimately overcome the disorder of geology, and this was how he himself viewed the interrelation between these two sciences. Geometry provided the mediation that could make the geology of the globe scientific. In the 1860s his enthusiasm for geometry extended to a defence of geometers as he vigorously supported Michel Chasles, the Professor of Higher Geometry at the Sorbonne, who in 1867 presented to the Académie des Sciences a large collection of manuscripts bought by him, unwittingly from a master forger. Proof that Pascal had anticipated Newton in the discovery of gravity was received with less chagrin than the eventual disclosure that among the letters purchased were those from Galileo, Cleopatra and Lazarus, all written in immaculate French.

Yet Elie de Beaumont was a great geologist, as a map-maker, and as an investigator of mineral veins. He was the first to suggest that mountains were not formed, as Hutton had believed, by the subterranean intrusion of large volumes of magma, but were the product of lateral crustal pinching and compression. Missing from his brief introduction to the English Geology had been the cautionary lessons against the dangers of theorising. His geometric Earth was standard teaching at the École des Mines for 30 years and his influence permeated all French Geology. With one foot on the solid ground of Geology and one foot on the unsteady pedestal of whole-Earth theories, through the accident of his journey into Geology, Elie de Beaumont served both as an example to geologists of the folly of grand Earth-plans and as an inspiration to others, who wished to address themselves to the greater scale of the Earth. For even after his crystal geometry had largely melted away, the underlying 'geometrical' concept of geological history survived as a fundamental, even structural, element of Geology.

For Elie de Beaumont offered a justification through his global revolutions for the construction of a global timescale, and many geologists were happy to accept the validation of this history without examining too closely its implications. Charles Lyell in his 1833 *Principles of Geology* vehemently condemned this synchronicity of world-wide mountain-building episodes, terming it 'an abuse of language'. However, the desire to make Geology scientific and universal countered Lyell's pleas for uniformitarian relativism. Already in England the great map-making geologist Adam Sedgwick had become a supporter of the mountain-building 'revolutions'. In the early 20th century, geologically based whole-Earth theories were to emerge that were little more than the fortification of the concept of correlation itself.

Elie de Beaumont's geometry lives on in isolated cultural relics. The belief that the orientation of structures identifies a particular age of mountain building is still received dogma amongst many geologists in England who rename the compass, calling E–W structures 'Hercynian', NE–SW structures 'Caledonian' and NW–SE structures 'Charnian', in the belief that an episode of mountain building is ruled by a geometry fallen out of the Books of Euclid. And then there are the mysterious workings of the zeitgeist. While Haussmann was serving his prefectural apprenticeship in a variety of rural assignments he took great interest in all aspects of the local geography and geology, later writing detailed accounts of these neighbourhoods. Elie de Beaumont's cobwebs, which permeated the standard geological maps, were an inspiration as to how disorder in the natural world should begin to gain symmetries. The crystal earth may have dissolved by the end of the 19th century but the geometry lives on in the radiations and perspectives of Paris, when Haussmann turned the crystalline state into a city.

Elie de Beaumont developed his innocent geometrical geology through the heyday of the science; the period from 1820 to 1860, before Charles Darwin's *On the origin of species*. Geology's success during this period had much to do with its failure to resolve certain controversies; in particular between itself and religion. Providential intercession in the running of the world was being beaten back, but only slowly stage by stage, and last of all from that most vexing question of the mutability of animal species and of man. What was the meaning of the development of life revealed in the fossils, which Lyell had stressed, displayed the deft hand of God through the whole of Earth history?

Yet for many Geology simply revealed the marvels of the Creator. An educated clergyman in his country living could turn his pastoral visits to fossil collecting. Geologising became a daytime justification for the traditional ambling walk through the fields and woods; as theologising occupied the fireside after dinner.

The battle between vulcanists and neptunists merged inexorably into one between catastrophists and uniformitarians; the factions were not the same but the argument sustained the same vital function within the developing science. Even after Lyell's plea for the uniformitarianism of past and present geological processes, the belief that the past held more glory and spectacle than the present sustained popular catastrophism throughout the Victorian era. Informed society listened avidly for the newest pronouncements of the geologists. The wave of unassailed success eventually had to break and this happened with extraordinary and unexpected suddenness at the end of the 1850s.

Charles Darwin's *On the origin of species* had many unforeseen repercussions. After Darwin, the innocence of the parson–geologist had become violated, and after Darwin, Geology became to be seen more as a force of darkness and materialism, working in the world to oppose the authority of the Church. Geology could not even benefit from the scientific maturity offered by Darwin, for *On the origin of species* had created a *new* science that claimed the fossil record on equal terms with that of living species. The conservative backlash to Evolution propelled Geology into an enervating argument with mathematicians over the Earth's antiquity. As if in sympathy with the spirit of the age it was at this time, as Geology suffered its first setbacks, that an attempt was made to reshape Elie de Beaumont's geometrical planet, by a global traveller, then entirely isolated from the European geological milieu, marooned on a Pacific island.

Elie de Beaumont had lived a dilemma. Wishing to produce a theory of the Earth, he had found that the more geological investigations he undertook, the weaker such a theory became as it was smothered in detail. The tension between his demand for geometrical symmetry and his actual observations as a map-maker ensured that throughout his life his geometrical pattern became more and more elaborate and ornamental. The moral was obvious: reduce the attention to geological complexity and concentrate on geometrical simplicity. The tetrahedron was not only the simplest geometrical solid, there was even a causal justification; the tetrahedron was the geometrical solid with the largest ratio of area relative to volume. Elie de Beaumont's force of contraction became dominant – the Earth had literally imploded.

William Lowthian Green was born in London in 1819, into a family of astronomical instrument manufacturers. After leading a life of travel and adventure, he arrived in Honolulu in 1856 where he quickly became a highly successful entrepreneur. In 1857 he sent to the *New Philosophy Journal* of Edinburgh the distillation of his thoughts on the

William Lowthian Green, 1819–1890
British vulcanologist and prime
minister

configuration of the Earth. Green had been studying maps, in particular maps of the continent of South America whose spine he had once traversed on horseback. The continents all taper to the south. Opposite every land mass there is an ocean. The apparent asymmetry of the Earth suggested an analogy with the tetrahedron. Around the southern hemisphere the alternation of continents and oceans could be modelled with six playing cards. The hexagon, Green explained, had been drawn towards a triangle; the deep-ocean centres are where the hexagon is pulled in; and the sharper angles form the mountain chains of the great land masses of Africa, South America and Australia.

Volcanoes have formed along the mid-ocean cracks between the faces. Fluctuating columns of magma within these volcanoes revealed the access to the deep Earth. The view from Hawaii was of a molten planet on which the crust was thin and insubstantial – a poor protection from the vast reservoir of red-hot magma stored not far below.

These first thoughts from Green were premature. They were, however, original, provocative and profound. He saw that volcanoes were the result of greater processes, and that volcanic activity was associated with the exact centres of the oceans. In 1875 he had his theories published as a book, *Vestiges of the molten globe*. His contracting Earth has now gained experimental imitation in the work of one 'Mr Fairbairn' who has tested the collapse of spherical bodies: 'Hollow glass spheres were reduced to fragments' – an unfortunate fate to prescribe for the Earth – but more suitably 'rubber spheres immersed in water tend towards the tetrahedron; and some organic forms like nuts exhibit shapes indicative of collapsing tetrahedral

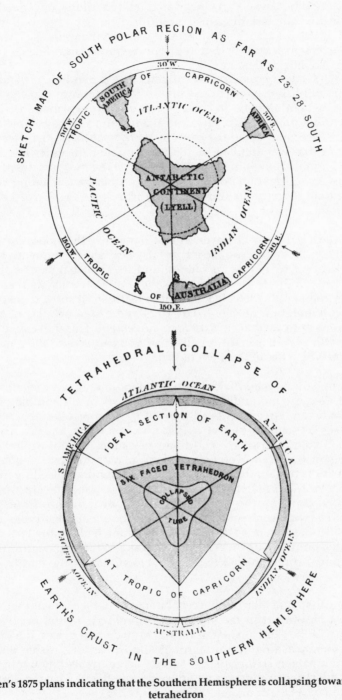

Green's 1875 plans indicating that the Southern Hemisphere is collapsing towards the tetrahedron

bodies'. This book was Part One of an intended trilogy, but Green was dismayed by the lack of critical attention. The two reviews that did appear in *Nature* and *The Athenaeum* were contemptuous. The only encouragement came in letters from Belgium and France, from followers of Elie de Beaumont.

In 1880 Green became prime minister of the Sandwich Islands. In 1882 his publisher wrote from London to inform him that the book was to be withdrawn. In 1882, without any explanation, Green resigned his premiership. His central ambition was to complete the combined second and third parts of *Vestiges of the molten globe*, an ambition that was fulfilled in 1887 and published under his direct supervision in Honolulu. Green had read all the most pertinent new theories emerging from Britain and North America, (including Fisher and Dutton, see pages 28 and 46) and he articulates a vision of a global tectonics:

> The earth is not yet, however, a dead planet, like the moon. On the contrary, all its features, the form of its oceans and continents, the direction of its mountain ranges, and its great bands of active volcanoes indicate life and motion. Not a motion truly, that can at every moment be detected, but these features all show it, just as the configuration of the "fixed" stars and of the nebulae of the heavens indicate that all is there in movement. This movement of the earth's crust, is in the broad sense, volcanic action, and it is an action which has left its own record.

Through the 30 years of observations made among these islands, the largest volcanic mountains on the planet, set in the middle of the greatest ocean, Green believed himself to have privileged access to processes taking place beneath the Earth's outer skin. Green's Earth is that of a volcano-dweller; Hawaiian volcanoes do not explode, do not demonstrate catastrophism, but gush with magma, frequently, copiously, as though the Earth is some haemophiliac, bleeding from a massive vessel of magma at the smallest irritation. The crust of the Earth is floating on this slippery base. His own visit to Chile gave Charles Darwin's comments on the 1835 earthquake special poignancy, and he quotes them in full. Two volcanoes had erupted at well-separated points along the Andean chain as a result of this earthquake. Darwin had contemplated this strange symptom of the Earth's convulsions.

> When first considering these phenomena, which prove that an actual movement in the subterranean volcanic matter occurred almost at the same instant of time at very distant places, the idea of water splashing up through holes in the ice of a frozen pool, when a person stamps on the surface came irresistibly to mind. The inference from it was obvious, namely that the land in Chile floated on a lake of molten stone.[8]

Green had little chance of communicating this perception to those who lived in the stable lands of North America and Europe. The desire to turn these simple perceptions into a global theory led him into geometry: besides the tendency toward the tetrahedron, there was also a crystal 'twin plane' passing close to the Equator – a great shear line of continuous rupture. Unlike Elie de Beaumont, Green was prepared to temper his geometrical demands according to at least some of the available evidence. The powerhouse of the geometry was contraction in alliance with rotation. The collapsing Pacific Ocean had pushed away the surrounding continents, forcing up a ring of disturbance. According to the Irish foundry manager and earthquake investigator Robert Mallet[9] the continents have lagged behind the rotation of the Earth because they are large and high; therefore the mountains have formed to the west of the continents. From Mallet, Green also finds that the general absence of continents in the southern hemisphere gives a differential lag: the south has therefore moved more towards the east, forming the great shear-zone twin plane that passes through the Caribbean, across the Atlantic, into the great Alpine mountain chain and off into the East Indies.

The global considerations, the tendency towards the tetrahedron and the westward flight of continents are intermixed with detailed descriptions of the Hawaiian volcanoes in eruption. Green notes the ubiquity of the mineral olivine in Hawaiian lavas, which makes him believe it has some special significance in the underworld from which the magma has emerged.

The effort in completing this, his life's work, had taken a heavy toll on his health, but as the work was being published he reaccepted the premiership. Several copies of Part Two were sent to leading scientists in England and to his few European supporters. For in France and Belgium Green's theories were already familiar. W. Prinz of the Royal Observatory in Brussels wrote of torsion, mirroring Green's twin plane, as being visible in other planets.[10] Recognition in his native England had to wait 10 years after Green's death when in 1899, J. W. Gregory produced a talk in defence of Green to the Royal Geographical Society.[11] In 1900, C. H. Hitchcock wrote an article about the man and his theory for *American Geologist*.[12] Two decades later many geologists had come to accept Green's tetrahedron as a fundamental principle that any theory of the Earth must be able to explain in order to be convincing.

The geometrical Earths of Elie de Beaumont and Green were the first attempts to grapple with a whole new scale of investigation. As Euclidean geometry had suddenly blossomed in the science of the crystals, why should it not so flourish in the science of the Earth? However, in place of these abstractions, these mixtures of insight and fantasy, new methods were emerging to provide information on the buried, hidden planet: the substance, the very stuffing of any whole-Earth theory. A new science of Earth Physics had emerged. It spoke in the language of Mathematics. Any serious Earth theorist had to listen.

The war between the geologists and the physicists

The concept of 'the depths of the earth' as Lyell had recognised in his list of 'prejudices', was easier to envisage than that of the whole-Earth because it played no tricks with perception but was a response to a simple question: what lay underneath? Yet as the bowels of the Earth could not be visited, nor mapped, it was of no particular concern to the little earth geologists. Deep into the earth and the whole-Earth were to them equally speculative, and therefore dangerous, territories in which to wander. Yet geologists had built up some requirements about this structure; most importantly that out of sight was out of mind. The crust of the Earth had to be thick enough to be self-supporting as Geology itself required no need to seek intellectual support in any other discipline. Lyell had admitted that the centre of the Earth was hot and that this heat powered mountain uplift, but he washed his hands of any necessity to contemplate how this was effected, merely requiring, in the name of Geology, that this heat supply should be constant and everlasting. Thus Geology made demands on the Earth in the manner that a tenant makes demands on his landlord. Kelvin even accused it of requiring perpetual motion.

The study of the internal structure of the Earth, which began around 1830, has sometimes been termed 'physical geology' but this title is misleading as it suggests a subdiscipline, when in practice the researchers shared little common ground. 'Earth Physics' is more appropriate as it was this study that provided the core of Geophysics, a science (see p 191) entirely distinct from the traditions and methods of Geology.

The first physical measurements to have been used in the modelling of the Earth's outer structure were those of heat. As their shafts had been blasted deeper, miners had formerly supposed that they might break into where Old Nick stalked the passageways that led straight to the fires of hell. By 1830 enough measurements of the increase of temperature with mine depth had been taken to extrapolate from the near-surface measurements and to find that at 80 km down the temperature would be hot enough to melt any stone. This fitted conveniently with some hypothetical internal ocean of molten rock that was the source of all volcanic magma. Whether volcanoes should be understood as open windows giving access to the depths of the Earth, or as accidents of nature, irrelevant to the overall design depended a lot on one's domicile. To an untravelled Englishman volcanoes were like foreign revolutions, an aberration in the essential solidity and stability of the globe. A traveller to Sicily or, like Green, a long-term resident of Hawaii, living next to the boiler, would come to view the Earth's stability as ephemeral; to see England as a God-given isle of stability on a dangerous planet.

However, the molten earth found sympathy with hell-raisers in a variety of traditions; with cosmogonists who saw in it a confirmation of the theory of Laplace, that the Earth was formed from the condensation of a gas cloud belched out of the sun; and for a majority of the geologists, whether, like Lyell, they required an internal heat engine or, like Elie de Beaumont, they sought an explanation for the confused pattern of mountain ranges in progressive Earth contraction. The opposition to the molten Earth was prompt in arrival: tidal forces surging through such a thin-membraned bubble of liquid would soon rip apart the crust. The argument was brought out and dusted down each time geologists reaffirmed their molten earth sympathies. According to the calculations, if the molten-earth was correct, everyone would be rocking up and down on the daily roller-coaster of magma tides, like ships in the London tide-pool.

The realisation that sufficient data were now available to allow mathematicians full sway to ally physical and chemical data with known earth properties led William Hopkins to begin working on these problems during the 1830s. Through his considerable influence as a Cambridge tutor to the most notable mathematicians of the mid-19th century (including William Thomson, later Lord Kelvin), Hopkins initiated the English school of Earth physics; a school that engaged the best mathematical minds of the age in pondering and calculating the problems. Lyell's influential, but coy, remarks about the constancy of Earth heat throughout geological time were an affront and a challenge to any physicist who cared to read the greatest work available in the popular science of Geology. Hopkins considered the nature of the problem.[13] First, he recognised the significance of convection in any heat transfer within a molten interior. Second, that the increase in the melting point likely with pressure would confound any simple extrapolation of the near-surface temperature gradient to some depth at which solid turned to liquid. Thus, he had to find a new source of information. The shifts in the orientation of the Earth's axis from the influence of the Moon and the Sun (the nutation and precession) were, he calculated, dependent on the proportion of the Earth that was solid, not liquid. The test proved to his own satisfaction that the crust of the Earth was solid down to at least 1000 miles. The volcanoes must feed off local subterranean magma lakes.

Hopkins then made a crucial and premature discovery that offered an open door into a whole new field of study. According to Fourier the temperature gradient should be inversely proportional to the thermal conductivity of the material involved. Hopkins compared the thermal conductivities of various rocks and found that this relationship did not hold. Hence he concluded 'at least part of our heat now existing in the superficial crust of our globe is due to superficial causes'. What was this additional source of heat? The open door was slammed shut; he used the information only to reinforce his view that projecting the near-surface temperature gradient deep into the Earth would produce great exaggerations. The door remained closed until the discovery of radioactivity. Hopkins died in 1866 before his 'superficial heat source' could be used to counter the claims of his mathematical protégé, William Thomson.

The dispute over the internal structure of the Earth was yet fully to implicate geologists who persisted in maintaining that as the interior of the Earth could never be 'known' in the way that the rocks of a quarry can, so all theories were of essence metaphysical. This pattern of unconcern became altered when Thomson, irritated as all physicists by Lyell's indifference to the new laws of thermodynamics, chose to launch an assault on Geology. If geologists were relatively unconcerned about the nature of the deep Earth, they were extremely sensitive upon the theme of time, believing that they alone were the guardians of earth-history.

By 1860 'fluidists', as the geologist supporters of the molten earth termed themselves, were suffering under a progressive and rapid process of solidification. Thomson[14] wished to exclude all liquid, apart from a few irrelevant puddles, from the planet. His justificatory argument went as follows: the phase change from liquid to solid ensures that rock is heavier than magma; therefore rock will sink as magma crystallises and so the Earth must have solidified from the centre outwards. As continents have not been known to turn turtle and founder into the Earth's molten abysses, so the planet must be solid. Fluidists countered the argument claiming that the natural analogy was with ice floating on water; an example considered by Kelvin and other physicists to be the misuse of an exception. Experimental information was yet to provide unambiguous results, and at least some crude observations had hinted that iron and other metals floated in their own liquids. The truth of the matter was not resolved until after the end of the 19th century by which time the shape of the greater dispute had itself become transformed.

Having provided himself with a solid Earth (*un oeuf dur*), Kelvin could subject it to a rigorous and simple mathematical analysis. In fact it was the possibility of achieving a mathematical analysis that provided the greatest reason for supposing that the Earth was solid. All alternatives were beyond the reach of the physics of heat transfer then available. Thus Kelvin's Earth was modelled as a 'red-hot cannonball in a water-filled bucket' planet. Kelvin's first target was not Lyell, who had talked of 'indefinite' or 'limitless' expanses of time, but Charles Darwin, whose recently published *On the origin of species* had spiked the romance of Geology and Religion. Darwin had spelt out lengths of time: 300 million years for the denudation of the Weald – the removal of a whaleback of sediments once thought to have run high into the sky, connecting the North and South Downs of south England and now eroded away to leave the sandy forests and vales of the mid-Sussex countryside. In defence of the reality of design in nature, and at the head of a crusade against the insane heresy of natural selection, Kelvin charged at this figure of 300 million years, preferring to dispose of the infant theory of evolution by starving it of time. Life was inconceivable without heat and light. So first there was a limitation on the Sun's age obtained from a knowledge of its annual energy loss; the only conceivable means of refuelling came through the collapse of flights of meteors onto its surface. 'It seems . . . most probable that the sun has not illuminated the earth for 100,000,000 years'. The Earth provided more direct measurements than the Sun and it was the Earth to which he returned for further restrictions on time. With a series of measurements and best assumptions about the starting conditions of the solid earth he arrived at an age of 98 million years which he bracketed in a range of 20 to 400 million years. Kelvin's third line of argument came from the retardation of the Earth's rotation caused by energy loss in the tides. Kelvin believed that only molten rocks can be deformed and therefore that the figure of the Earth – the flattening at the poles – had developed when the Earth was still soft. Thus a knowledge of the slowing down of the speed of rotation from tidal retardation could place a limit on when the Earth had frozen. The numbers of this last estimate were too uncertain, but Kelvin could see that they were in broad agreement with his Sun's energy and earth-heat calculations.

From 1861 until the end of the century geologists found themselves unwillingly locked in combat with the Earth physicists led by Kelvin. The problem was far more severe than a mere restriction on geological time, for Kelvin demonstrated that the world of Earth and Sun was engaged in progressive and irreversible cooling, on a timescale sufficiently rapid to play havoc with uniformitarian notions of constancy throughout geological history. This time-war dominated the larger-scale, theoretical and external concerns of Geology. Many geologists returned to simple field observations in the hope that the storm would blow itself out. The intuitive understanding of a dynamic earth held by geologists was matched against a 'rigorous' physical model; it was cannonballs against fieldwork. Of the three defences available to those who opposed Kelvin (additional sources of heat, incorrect extrapolation of physical parameters to high pressures and demands for a non-solid Earth) only the last possessed a rhetorical simplicity that could hope to challenge the cannonball.

The defenders of Geology

In the last quarter of the 19th century a new champion emerged to fight with mathematical weapons on behalf of the besieged fluidists. The Reverend Osmond Fisher was born in 1817 and grew up in an age when Geology provided the natural language with which to investigate the countryside. The professional had yet to cast the shadow of the amateur. Fisher was a parson naturalist who became ordained only after reading Mathematics at Cambridge University, where he graduated four years ahead of Kelvin. He sustained himself through a series of country and college livings with much time spent in the chalkland and heaths of Dorset and Wiltshire, before being made a fellow of Jesus College, Cambridge and, in 1867, rector at the village of Harlton close to Cambridge. Earth theories had, he claimed, presented

Rev. Osmond Fisher, 1817–1914
British mathematician and geologist

themselves to him when, in the winter of his final year as an undergraduate, he had seen the ice, in one of the locks of the River Cam,
cracked, ridged and frozen again. Without the benefit of Elie de
Beaumont's geometry he had experienced the contracting earth. 'In
his delight at the discovery he forthwith vaulted over a fivebar gate'.
Soon after the death of his wife in 1867 he began once again to
meditate upon the Earth. During 1868 he presented a series of lectures
that supported both the theory of contraction and Kelvin's solid Earth.
However, between June and December 1873 (the date of two papers on
the subject) Fisher became a born-again fluidist. The June paper[15]
defended the solid-Earth contraction through attempting to calculate
the height of mountain ranges produced from simple shrinkage of the
solid Earth. By December[16] he was advocating a part-fluid Earth,
ostensibly because new calculations showed this solid-Earth model to
be incapable of producing a sufficient range of elevations. No simple
calculation could have provided so marked a shift, and it is clear that
Fisher, newly self-confident, was simply returning to the model with
which he had grown up and which had only been abandoned because
he found it, in middle life, to have become opposed by a new
orthodoxy. The confidence persisted. In 1880 he began writing *The
physics of the Earth's crust*, the first book ever to build mathematical
reasoning on a geological foundation. The work was published at the
end of 1881 and, despite some hostility from the physicists, gained a
considerable sale. Fisher was now 64, yet the work appears like
a doctoral thesis that was to determine the direction of all his
subsequent interests.

The book first attacks the solidist contraction hypothesis for its
failure to explain the surface elevations and then constructs in its place
the thin-crusted Earth over a fluid substratum within which there was
a large solid nucleus. Fisher deflects the established attack against a

28

liquid Earth claiming that the absence of tides was explained by the dissolved steam in the magma (as evidenced in volcanic eruptions) emerging and redissolving to provide perfect shock absorption for the overlying crust. Fisher's fluid was, most importantly, undergoing convection. Although the Earth was engaged in long-term cooling and contraction, the convective flux was providing horizontal drag, pushing up mountain ranges and opening rifts. The lava lakes of Kilauea revealed slabs of congealed magma sliding over the molten rock below. The convection kept the underside of the Earth's crust at a constant temperature and therefore allowed a thin-crusted Earth to be of sufficient antiquity for geologists to dictate what draughts they would require from the wells of Time. The ice in the frozen lock; the lava-floes; these, not cannonballs, provided models of the Earth.

In 1909 at the age of 92 Fisher was still writing; 'On convection currents in the Earth's interior'.[17] Undeterred by the hostility of the mathematicians, who preferred their calculations to be based on available physical models of explanation, Fisher considered that Geology was the more fundamental science. Spanning all the intellectual developments of the 19th century, he died before his central belief in convection could be summoned to support the new theories of continental drift.

The second substantial attack to be mounted on Kelvin, before the discoveries of radioactivity had rewritten the terms of the argument, came from an American, Thomas Chrowder Chamberlin. Chamberlin, born in 1843 into the wide prairie landscapes of the American mid-West, was not to know what geologists in England considered intellectual etiquette because he was self-taught; one of the great American individualists who became educated through teaching others. The quality of the eternal pedagogue built for him a fortress of self-certainty. A self-taught person has the extraordinary capacity to

**Thomas Chrowder Chamberlin,
1843–1928**
American geologist and cosmologist

reinvent the historical development of a science, to approach it from new directions, to assess each problem in turn and to arrive at new solutions. The challenges and the victories encountered in the journey of self-discovery made him unassailable. Thus Chamberlin was to become an emperor in his science.

After graduating from his local Wisconsin college at Beloit he undertook teaching, preaching and geology until in 1881 he became chief of the US Geological Survey's glacial division. In 1892 he left his home state to become head of the Geology Department at the newly created University of Chicago. The faculty intended to serve in his geological cabinet was hand-picked to contain the brightest and best. To teach the world he immediately initiated and edited a strong mouthpiece magazine: the *Journal of Geology*. The empire was now consolidated; the vision could begin to roam free.

Mapping the geology of Wisconsin had none of the glamour of the wild-west surveys but it did serve to concentrate the mind. The landscape had been crumpled, dimpled and scraped by an ice sheet – this was already known. The sleeper had fled, leaving the bedclothes in disarray. The evidence was everywhere. Chamberlin's success was to have shown that there was not one but a series of advances and retreats from the great ice tide. Having spent a lifetime mapping the effects of the ice sheet Chamberlin was only brought face to face with the sleeper in 1894 when he accompanied the US explorer Peary on an expedition to Greenland. The silent grandeur of the glaciers pouring off that mountainous subcontinent into the sea, impressed on him for the first time the horrendous scale implicated in a whole North American continent of ice, which had created the pleasant grassy knolls and meres of his beloved Wisconsin.

Already as a teacher he had been moving towards philosophy; certain of the truth towards which science was always ascending, Chamberlin wrote many articles on his own heuristic method, caught in the paradox of the self-taught, wishing to teach others how they too could be self-taught. Chamberlin believed that Geology could provide the richest intellectual environment within which such an education could take place. 'The earth sciences are not purely physical science; they concern themselves with life and with mentality as well as rocks'.[18] His Geology Department was to be as broad and as strong as those of the renowned Chicago Schools, Philosophy and Social Thought.

In 1899 Chamberlin was triggered into action after reading an account of a speech delivered by Kelvin. He wrote a reply[19] denying certain limitations on the declining heat of the Earth, which had become evident to him through living in, and studying, a landscape that had once been far colder than it was now. There had also, deep in the geological past, been similar ice ages in South America. Scorched by the harsh Mid-West summer sun Chamberlin was inspired to write a sentence that was to create his reputation as a seer: 'It is not improbable that (atoms) are complex organisations and the seats of enormous energies'. Radioactivity had still a few years to be identified as a heat source, just long enough for the words to gain considerable resonance and thereby to reinforce Chamberlin's sense of authority and infallibility.

With the consolidation of his empire of the Earth, Chamberlin traversed the boundary into Astronomy and Cosmology, the first respected scientist of the 19th century to have made this transition. He even found a name for this bridged science: 'Geocosmology'. Chamberlin moved out of Geology for the same reason Napoleon moved out of France; as a demonstration and confirmation of power in his home base. This image is born out by Chamberlin's own sense of the interrelations of the sciences: Astronomy was no more than 'the foreign department' of that most synthetic of all sciences, Geology.[20]

To investigate this alien territory Chamberlin needed a native speaker. In 1898 he journeyed to the Chicago University Department of Astronomy and there found a mathematically accomplished, 26-year-old Astronomy assistant named Forest Ray Moulton whom he invited to participate in an investigation of the Earth's origin. After a total solar eclipse in 1900, Moulton and Chamberlin scrutinised photographs of the Sun's corona that revealed massive surface eruptions. Such eruptions had been proposed by the American mathematician and astronomer Daniel Kirkwood, who from 1856 had gained the chair of mathematics at Indiana University, where for 30 years he continued his research into the origins and nature of the solar system. The clues were to be found, Kirkwood thought, not through observing the all too familiar planets, but instead in the proletariat of the solar system: the asteroids, comets and meteoritic bodies. From 1869 he attempted to build a new theory to replace the Laplacian 'condensation from a gas-cloud' for the origin of all the bodies of the solar system, large and small, in which the planets had accumulated from arcs of matter expelled from the shrinking Sun.

Through Moulton, Chamberlin found additional mathematical arguments to justify his scepticism about the Laplacian theory. Gravitational attraction of the atoms in a gas was insufficient to aggregate the mass. Kirkwood's large flocks of asteroids roaming the space between Jupiter and Mars provided the material for a new model. Chamberlin's conversion was complete by 1904 when he started to publish his 'planetissimal theory'.[21] Inevitably the asteroids were not isolated incidents in the history of the world – they provided the very building blocks of creation.

The Earth was contracting, not from cooling but from a 'down-crowding process' whereby the untidy pile of asteroids, out of which the Earth was composed, settled like some cosmic rubbish tip. Most significantly therefore, in the instant before Kelvin was about to encounter hubris, the first victim of radioactivity, Chamberlin created a contraction theory that was heat-independent and could thus survive unscathed the discoveries of radioactive heat generation in the rocks. Contraction also ceased to be part of some overall world decline; downcrowding had none of the apocalyptic evocations of Kelvin's Earth sliding headlong into the last eternal winter.

The magnitude of the collapse could be found by comparing the Earth with its neighbours. The low-density Moon must be as honeycombed as Gruyere. With the Earth and the Moon built from the same material, the denser Earth must have shrunk by a vertical distance of at least 725 miles (1200 km). That this collapse took place in great episodes of upheaval was Chamberlin's view and one that he democratically transferred, in 1909, to 'the view that prevails, I think, among American geologists'.[22] These episodes of collapse and subsequent rises in sea level as the oceans became infilled with sediment, offered world-wide correlation. As Chamberlin had implicated the nation in his theories, to disagree with him was to countenance that worst of all American sins, antipatriotism; for after all 'Correlation by base-levels is one of the triumphs of American geology'.[22] Reading Chamberlin can be like reading Lenin: the self-assurance, the bulky neologisms, the didactic, insistent patter of 'truths' cleaned of self-doubt or self-reflection, the flow of the sentences, building statements on statements without a nuance of hesitation, all is part of a totality of views that permeate all endeavours. Chamberlin was a great ideologue, preaching the American creed: that only individualism led to enlightenment. He retired from his university position at the age of 75 in 1918 but continued as editor of the *Journal of Geology*, his remaining power base. When he died in 1928 his friend, the geologist Bailey Willis, in some overstatement characteristic of the man himself, provided an epitaph:[23] 'Aristotle, 322 BC; Copernicus AD 1543; Galileo, 1642; Newton, 1727; Laplace, 1827; Darwin, 1882; Chamberlin, 1928.' He had taught his fellow geologists well.

At the end of the 19th century the Austrian geologist Eduard Suess began compiling his massive compendious *Das Antlitz der Erde*, a celebration of the great success of the science of geology that in only 80 years had provided the key to mapping the rocks of the whole-Earth. The title of the book was chosen as an honest statement that Geology was not the science of the Earth – merely of its face. Suess's position in the greater history is pivotal; all subsequent developments in the investigation of mountain ranges were to orientate themselves with regard to him. He wrote of the findings of the old world of ideas but defined the outlines of the new. His four volumes provided the culmination of the geological method. They push at the boundaries of scale and scope imposed upon Geology, yet never break free. Disbelieving of isostasy he affirmed the geologist's credo: 'it is with extreme distrust that the geologist regards all attempts to apply the exact methods of mathematics.' No longer possessed of Elie de Beaumont's confidence at having comprehended the macroscopic behaviour of the Earth, discounting the synchronous catastrophic episodes of mountain building, Suess saw geological correlation only being possible through changes in sea level; an idea that he passed on to Chamberlin.

Eduard Suess, 1831–1914
Austrian geologist

Throughout the 19th century, whatever the arguments with the Earth physicists over Earth structures and Earth age, there was a general consensus that the Earth was contracting, and that the crust of the Earth was somehow detached from this internal retreat. With an aristocratic disconcern about the underlying substance of the Earth, Geology was involved in detailing the Earth's progressive morbidity; writing its Requiem Mass. 'What we are witnessing' writes Suess with Nietzschian portentousness 'is the collapse of the World'.[24] In 1878, at the first International Geological Congress in Paris, the French had presented their contracting Earth in the form of an experimental display. As B. de Chancourtois pulled the cork from his large, wax-covered rubber balloon, the participants were privileged to see, before their very eyes, curved mountain ranges rising out of the crust as the wax ocean floor became encircled by ugly welts. The rubber balloon seemed an ignoble model for man's planet home and in pursuit of a more poetic image, Suess, echoing the hermetic philosopher Giordano Bruno some three centuries earlier, proposed that the Earth was an apple. Not a fresh fruit for tempting Eve, nor a ripe windfall to inspire Newton, but an apple in decay. Shrivelled, desiccated, its skin crumpling as the interior retreated. The mountain ranges were, as if to counter the Romantics, not a symbol of vitality but wrinkles of senility. *Un pomme de terre.*

3

New world

A mental Geography

The 19th century is littered with clues, inconsistencies and observations that were, within the first decade of the 20th century, to
become harnessed to support the hypothesis of continental drift. They
are like dream-images awaiting ordering. Disconnected items sometimes existing out of context, sometimes triggering associations. Maps
of rearranged continents were a particularly powerful stimulus as they
could linger in the unconscious, independent of their original source.

The first such map has an unlikely location within Lyell's *Principles
of geology*. The map is presented as a fantasy. Lyell is attempting to
prove his uniformity of past and present Earths against all anticipated
opposition. The discovery of coal deposits (fossilised tropical forests)
on the icebound island of Disco off the western coast of Greenland
requires some explanation. Lyell presents his argument: climate is
very dependent on the distribution of land and sea, an influence that
can be found from the present disparities between the climates of the
Northern and Southern Hemispheres. The maps are playful illustra-
tions of how the present proportions of land and sea could be
rearranged into the extremes of an equatorial or a polar distribution.

Lyell's maps are innocent, but their significance is altered entirely
when viewed anew after having read the American geologist Dana's
remarks, made in 1846, about the permanency of ocean basins.[1]
Continents stay continents; oceans stay oceans – no sea-change can
effect a complete transformation. Therefore Lyell's argument falls;
unless the continents themselves can move. Dana's immutable oceans
were to prove most controversial as their acceptance seemed to require
extraordinary explanations for the similarities discovered from one
isolated continent to another. Yet it is significant for Lyell's maps that
when arguments about continental drift did finally break into the
open, they came through an alliance with Climatology.

The recognition that the two sides of the Atlantic mimic one another
is as old as maps. By the end of the 16th century the cartographers had

MAPS showing the position of LAND and SEA which might produce the Extremes of HEAT and COLD in the Climates of the GLOBE.

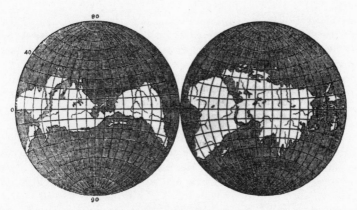

Extreme of Heat.

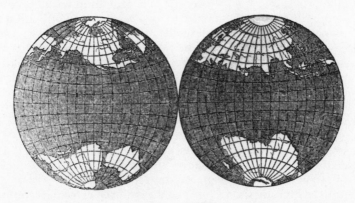

Extreme of Cold.

Lyell's 1841 re-arranged continents from *Principles of geology*, to show how the distribution of land and sea could affect climate

given the continents their extravagant outlines. Featureless apart from these serrated coastlines, it was possible for any vicarious explorer poring over the charts to see the similarity of forms. Anyone? It is the child who sees faces in rocks, witches in trees, castles in the clouds. The great bulge of Africa squeezed into the embrace of the Americas. For some it was the 'S' shape of the Atlantic that attracted the eye. Was it a hallmark of the Creator? In 1801 Baron Alexander von Humboldt claimed that the parallel coasts were river banks, and that the ocean was an enormous erosion valley, cut through the transecting mountain ranges as flood waters pouring in from the south had burst over the Earth.[2]

Before Humboldt, at a time when all history was to be found only in written texts, the origins of the Atlantic were also the story of Atlantis. Plato's brief second-hand report of a nation of advanced civilisation and prosperity, destroyed a long time before in a cataclysm and subsequent inundation, was central in any elaboration of the biblical account of Creation. In 1620 Francis Bacon[3] claimed that the Americas were the remnant of Atlantis, and also mused on the similarity of forms of Africa and South America.

On his death in 1506, after four successful voyages to the west, Columbus still refused to believe that he had discovered anything other than the eastern fringes of Asia. This was a poor beginning for a pair of new continents: America was some European hallucination through which one might resolutely sail all the way to China. To the Puritan settlers their 'promised land' was in danger of existing in limbo, without the blessing of prophecy. That America was insubstantial, that the world had existed for so long without knowledge of these vast western lands, has persisted ever since as 'the paradox of the Americas'. This paradox takes the following form: culturally, spiritually, right into the 20th century the affiliation with Europe has remained at the core of American society. Yet this affiliation was accompanied by a distance that fuelled commercial and, later, cultural rivalry. These counterviews of the location of America – the tension between sameness and difference – are those provided by the mirror. At the heart of this paradox of location, in the line of the mirror, lay the Atlantic Ocean.

Towards the end of the 17th century there emerged a desire to make the Flood in Genesis compatible with rational explanation. The new discoveries of science and the new era of Cartesian philosophy had threatened to undermine the Bible, and now there was a need to show that all these changes could hope to elucidate, not contradict the biblical account. There were many ingredients of the story of Noah that required investigation. How had the children of Noah repopulated the barren Earth? How had the animals been stored in such cramped quarters? What of the amphibians, the birds and fish? Where had the water come from, and disappeared to? All these questions demanded explanations; the discovery of America had served to complicate some of the simple solutions first proposed. As expeditions into the interior of the Western continents revealed the presence of more and more new animals and tribes, the problem became still more perplexing. The American Indians had no traditions of navigation. Milius[4] in 1667 discussed the difficulty of freshwater fish and heavy-winged birds crossing from the Ark to the Americas. Men could have transported the domestic animals by boat but would have been crazy to have imported 'lions, tigers, bears, dragons and serpents.' His suggestion was that the water had only been 15 cubits deep over the Asian mountains, which left the taller American mountains uncovered. Hornius[5] in 1669 considered that the animals had crossed in winter at which time the northern seas were frozen. Kircher[6] in 1675 offers a variety of explanations: 'that the animals swam from island to island, crossed by isthmuses, were conveyed in boats, or escaped from ships that were transporting them to ducal zoos.' Hale[7] in 1677 believed that there had been a single land bridge, subsequently washed away. This idea of land bridges had also impressed Kircher, who even prints a map showing such causeways running from Brazil to Greenland and from southern California to Chile. At the time of the Flood the world's islands were formed: Madagascar had formerly been part of Africa and the British Isles of Europe. Such games of geographical rearrangement had become so popular in Stockholm that in 1699 the Professor of Geometry was prevailed upon to give a public lecture[8] to detail how at the time of the Flood all the land bridges were broken and all the volcanoes snuffed out. The range of explanations used to explain the journey of the animals after they had disembarked, was to be used in an identical form 200 years later.

It was inevitable that within this fertile and popular desire to give more substance to the Flood, explanations would be provided that took still greater liberties with the positions of the land masses. In 1666 François Placet, a French Jesuit moralist, claimed that the present continents of the Earth were part of a much greater land area that became divided in the Flood. America had formed itself either through the agglomeration of floating islands or as a Western counterbalance of emergence to compensate for the collapse of the North Atlantic continent: Atlantis.[9] This was a European view, like that of Humboldt, in which America had always been a distant land but, since the inconsequential middle ground had been eroded, now marooned by an ocean. In 1756 a German professor of Theology in the University of Königsberg, Theodor Christoph Lilienthal, wrote an exegesis of the biblical flood[10] in which he interprets a biblical reference to the 'division' of the Earth as indicating a split of the continents, 'made likely by the fact that the opposite, though by the sea separated, coasts of many countries have a congruent shape so that they would almost fill up each other in case they would stand side by side; for example the southern part of America and Africa.'

In 1858 an Italian-American traveller, Antonio Snider-Pellegrini, living in Paris, wrote his major work *La Création et ses mystères dévoilés*, intermixing scientific observations with the traditions of the late 17th century, rewriting the days of Creation as epochs in the history of the world. On 'Day' Six, in the afternoon following the busy morning of the Creation of Man, there is the formation of America. A sudden outburst splintered the newly congealed crust sending fragments of the original continent slithering across the globe, like shards of ice across a frozen pond. This catastrophe caused an enormous surge of water, pouring only over the continents to the *east* of the Atlantic, which launched Noah's zoological lifeboat. 'If Franklin had lived he would share my opinion in the origin of America' claimed Snider-Pellegrini. (Franklin might well: in 1782 echoing Robert Burnet's theory of the origin of the Earth's features, he imagined that 'the surface of the globe would be a shell, capable of being broken and disordered by the violent movements of the fluid on which it rested'.[11])

The underlying motive for Snider-Pellegrini's work is to be found in appendices on 'the moral man' and 'the disparity between civilised and barbarous races'. While Europeans and European-Americans share ancestry that included both Adam and Noah, the native Americans, who survived the Flood unscathed, were descended only from Adam. The first 19th-century demonstration of continental drift was therefore no more than an attempt to justify the persecution of the racially impure American Indians.

The mid-19th century was not a propitious time to launch myths. Science was the new source of histories and explanations and one year after *La Création et ses mystères dévoilés* had appeared, *On the origin of species* was challenging many of those assumptions that a myth is intended to assure. By all expectations Snider-Pellegrini's story of Creation should have sunk without trace. However, something remarkable and totally unexpected happened to the book. It was discovered within a few years of publication by an English scientific populariser: John Henry Pepper. In 1861 Pepper brought out a best-selling text, *The playbook of metals*, which begins with a discussion of the origin of coal. The replacement of land by sea, necessary to explain the location of the Coal Measures, leads into a presentation of Snider-Pellegrini's two globes, showing the Atlantic before and after separation. The theory is qualified only by the statement that 'M. Snider supposes' and Snider's remarks are immediately followed by mention of Lyell and Darwin. The American has been elected a scientist by the company he shares. Most significantly, Pepper strips Snider-Pellegrini's theory of all mention of catastrophe or the sixth epoch and instead provides no clue as to the period of time over which such separation took place. The maps tell their own history. At the head of the double page containing Snider-Pellegrini's 'Arrangement of the land after the separation' there is written with an eerie predestination 'The drift theory'; a name that has been plucked from 60 years into the future; yet here a reference to the belief that timber logs had formerly floated to the coasts of England to be turned into coal.

Coexisting worlds

The very frequency of reproduction provided the world map of the continents with conceptual permanence. The first attempt to redraw the map of the world came not from Geography but Zoology. In 1856 the bird and insect collector, traveller and philosopher, Alfred Russel Wallace, had been island-hopping from the west down the Malay

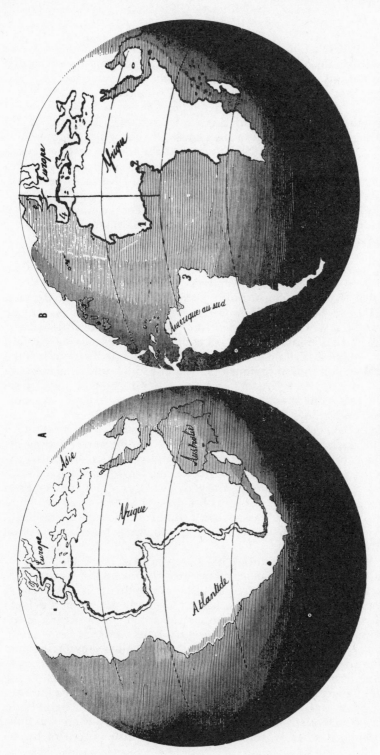

Antonio Snider Pellegrini's 1857 globes, 'avant' and 'après la séparation' – the opening of the Atlantic causes the biblical deluge

Archipelago when he traversed the 30 km of sea that separate the twin volcanic islands of Bali from Lombok, and found himself surrounded by an entirely new fauna: in place of the green woodpeckers and barbets that he had collected on all the islands beyond the Malay peninsula, there were sulphur-crested cockatoos, honey-suckers and brush turkeys squatting on their brick-red eggs laid atop giant mounds of vegetation. These alien species he recognised as creatures of Australia. In January 1858 he wrote 'there are two distinct faunas rigidly circumscribed, which differ as much as those of South America and Africa, and more than those of Europe and North America. Yet there is nothing on the map or on the face of the islands to mark their limits. The boundary line often passes between islands closer than others in the same group.'[12] Wallace had found intimations that were to lead to the mapping of a separate geography, coexisting with, and sometimes contradicting, the mapped geography of the Earth. In this new geography the 30 km between Bali and Lombok were further than the 3000 between Spain and America.

In 1858 the ornithologist Philip Lutley Sclater presented a paper at the Linnean Society of London in which he detailed a scheme whereby the world was divided into six regions according to similarities and differences in the bird population.[13] Wallace was inspired to undertake a further examination of the boundary between the Australian and Indian regions that Sclater had left undefined, and in 1859 wrote a paper[14] tracing the line separating these two faunal provinces – a line that now bears his name. It was to prove the most striking and sudden faunal boundary on the planet.

Sclater believed that the faunal regions represented the locations of specific acts of Creation; but Wallace, who had been developing his own system of evolution independent of Darwin, perceived that the pattern revealed the interrelations of the land masses, and in particular their past interconnections. As the geology of this region had not been mapped, Wallace could not rely on a geological history on which to locate his demarcation. Instead, in a unique inversion, he attempted to unravel the geological history from the evidence of the modern faunal distributions.

In 1876 Wallace produced his compendious two volume, *Geographical distribution of animals* that was to become the main source work for the new science of Zoogeography. In it he discussed a number of the problems antithetical to those of the Malay Archipelago; problems of the similarity of faunas separated by an ocean; of the marsupials, chyelids and the tree frog *Hyla* that were common to both Australia and South America. Some of the most convincing and extraordinary similarities had been described in 1864 by Sclater[15] in an attempt to comprehend the mysterious island of Madagascar.

Although located only 500 km from the coast of Africa, the fauna of Madagascar has very few similarities with its western neighbour. There are no monkeys, apes, baboons, lions, leopards, hyenas, zebras, rhinoceros, elephants, giraffes or antelopes. Instead, out of the 65 known mammalian species, more than half are species of lemurs; primitive monkeys, whose tree-bound nocturnal habits and ghostly demeanour had merited their Latin name for the souls of the departed. If there were any faunal connections to be made then they were with distant India. Sclater coined the name 'Lemuria' to include Madagascar with India and suggested that a drowned continent had once included both these lands. Wallace, too, toyed with Lemuria, only later to dismiss the sinking of continents as unrealistic. However, as Wallace was the bridge between Zoogeography and the British spiritualist movement, so Lemuria was to gain a new significance. The drowned kingdom of the dead, as its literal meaning implied, became the substantiated location of an arcane lost civilisation whose inhabitants could be summoned and interrogated through the Ouija board and the drawing room seance. These lost continents of Zoogeography were no more substantial than the lost civilisations of the occult.

Yet while a zoologist could walk among the wildlife, gaining an impression of the feel, the very harmony and continuity of biology, noting how it was the whole pattern of species that varied from one continent to another, the reconstruction of past wildlifes was achieved by the collection of individual fossils. Zoology was extrovert where Palaeontology was painstaking. Zoogeography was the product of walks in a variety of jungles, while Palaeo-zoogeography required browsing in scientific literature. The pattern of past life was beginning to reveal a geography still more at odds with the modern map of the continents. This story of the fossils trickled out of Geology to reinvigorate the myth of Atlantis. After an American populist politician, Ignatius Loyola Donelly, had in 1882 employed the new information to bolster a best-selling book on Atlantis, the British prime minister Gladstone was so enthralled that he attempted to steer his ministers into authorising an expedition to search for the underwater realm.

By 1885 the fossil correlations had become sufficiently numerous for a German zoologist, Melchior Neumayr, to draw a world map of Zoogeography for the Jurassic geological period.[16] Europe is reduced to a few inkdrop islands, North America has expanded its empire to overrun Greenland and Iceland, China is linked with Australia, and most significantly, the 'Asia' of the Jurassic is a vast land mass, the 'Brasilian-Ethiopian Continent', which includes the whole of Africa and South America along with the intervening South Atlantic Ocean. For the first time the view has shifted to the antipodes; the paradox of the Americas has become transferred from the North Atlantic to the South.

Within the boundary between the Permian and Carboniferous geological periods there was evidence of a great glaciation over Africa, South America and India. Across these same continents, and also into Australia, from the same general strata, there were the strikingly elegant fossils of a fern, *Glossopteris*, that seemed to unite these continents in a common history. The type-locality of this fossil was in

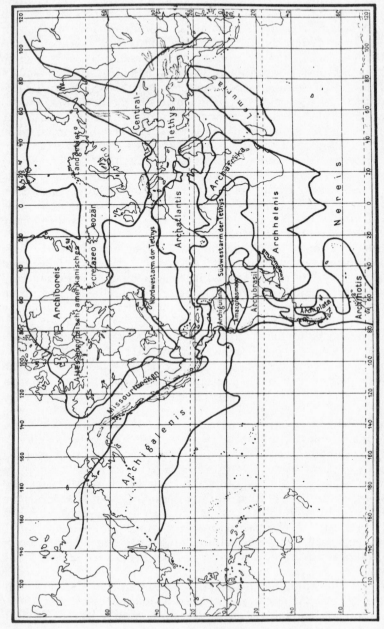

H. von Ihering (1927) constructs three land bridges (Archiboreis, Archatlantis and Archhelenis) across the Atlantic during the Eocene

India in a region called Gondwana – the land of the Gond peoples. Suess, in an innocent tautology, named the continent that had once united these widely separated places 'Gondwanaland'. At the same geological period, bowing to the primacy and authority of the recast mythology, Suess termed the matching Northern continent 'Atlantis'. Between Atlantis and Gondwanaland there lay as a great dividing stream, the Sea of Tethys; a name taken from Greek mythology to emphasise the new mythology of geological history – Tethys was the sea-nymph wife of Oceanus.

Suess never produced maps of his past continents, thus their status was abstract and immaterial. Continuous land masses were needed to justify the homologies of plants and animals, yet Suess's metaphysical continents had ultimately to coexist with physical continents of rock. The reason that Suess did not dare to draw his Gondwanaland was very simple: it was so huge – a great blanket of land wrapped around the Southern Hemisphere extending from Tierra del Fuego to the Great Barrier Reef and incorporating two whole oceans, the Indian and the South Atlantic. The conflict between a zoogeography and a geography, the conflict that had so baffled Wallace, remained unresolved and yet served as a great tension – a knot of problems that could not be unravelled.

The most economical solution to the problem of reconciling past and modern zoogeographies came from the construction of land bridges. These were substantial enough to explain all biological enigmas, and as slim and ephemeral as was required to avoid too overt a contradiction of any supposed physical law. They were the reification of explanation itself; the perfect compromise. Because the zoogeographic world was itself an abstraction they provided causeways of suitable irreality. Suess's reticence about his reconstructed continents naturally led to some more specific recommendations. In 1907 the German geologist Hermann von Ihering[17] reduced the oceanic sections of Gondwanaland to a fragile skeleton. In the Eocene geological period (part of the Tertiary) he spanned the Atlantic with three such bridges, but reserved his most stupendous meta-tectonic engineering to a great causeway across the northern Pacific. Land bridges were to provide the geologist's chief alternative to drifting continents. The Central American isthmus provided the perfect modern example.

In later years the American geologist Bailey Willis[18] was to champion a theory in which orogenic forces originate beneath the oceans to force ridges of oceanic crust to break surface. For a short while the animals scuttle to and fro, redistributing the species, before the inevitable collapse takes place, sending the land bridge diving back down to become absorbed once again into the ocean floor. Such a model was successful at explaining why such features have so mysteriously vanished but again was completely at odds with either the permanence of the ocean basins or Geology's most cherished philosophy of history. For if it was possible to squeeze the dense ocean crust to the surface, there would be a massive increase in measured gravity; a feature found over no possible analogous isthmuses such as Panama. Thus land bridges could only be saved by the claim that there were no modern examples; an expedient heresy to uniformitarianism. Given a choice most geologists preferred to dismiss the American argument for ocean permanence.

Balancing the crust

The crisis was gradually developing and would eventually lead to some kind of catharsis. The crisis existed because two strands of geological theories, those associated with the permanence of ocean basins and those demanding that oceans had in the past been as land between the continents, were irreconcilable. The crisis took time to develop, not because these two positions were altering, but because they were becoming better established. For while geologists were being made familiar with the fossil parallels from one continent to another, outside America they were slow to perceive the full significance of Dana's dictum about the permanence of continents and oceans, in particular the coherent conceptual framework that locked this permanence in with the formation of mountains as revealed by the remarkable Clarence Dutton.

To put Dutton's contribution in perspective it is necessary to revisit an earlier debate about the variation of gravity at the Earth's surface. Between 1838 and 1843, as part of the Great Trigonometric Survey of India, intended to enmesh the subcontinent with a national grid, the Venerable John Henry Pratt, Archdeacon of Calcutta, found that a plumbline used through the Ganges Valley, within sight of the great Himalayan chain, suffered far less offset from the vertical, due to the gravitational attraction of the mountains, than had been predicted. The deflection had shown up when the angles between one station and the next were measured with respect to the plumbline's 'vertical' and it was in this form that Pratt presented his results in London in December 1854.[19] The meridian of longitude over India, explained Pratt, must be more tightly curved than over the rest of the planet. Within a month, the Astronomer Royal, George Airy, presented to the Royal Society of London an improved explanation of Pratt's results.[20] As it seemed highly unlikely that the Earth's crust was itself strong enough to support mountains, the mountains must have deep roots of lower density material that extend down into some higher density medium. The crust must be floating in a state of equilibrium. The underlying substratum, termed 'lava' by Airy, was solid but yielding in the manner of the rock-walls of deep mines.

Four years later, in the year of *On the origin of species*, Pratt,[21] stung by the speed and originality with which Airy had reworked his, Pratt's, observations, produced a refutation that was entirely compatible with the new orthodoxy of the physicists then undergoing rigidification. The crust according to Hopkins was at least 1000 miles (1600 km) thick.

Airy's thin crust floating in lava was, therefore, riddled with inconsistencies: if such an underlying liquid were of the same composition as the crust it would prove lighter, not heavier. Thick roots for mountains would correspondingly imply thin crust for the oceans, 'leading to a law of varying thickness which no process of cooling could have produced'. Thus Airey's scheme, of the crust of the Earth as logs of differing thickness floating in a common pond, became infilled by the contraction hypothesis. The new orthodoxy required that all explanations should be tailored to suit its requirements: in Pratt's scheme the differences in elevation were from density contrasts, like the constituents of a plum-pudding, buried as 'hidden causes' within the outer shell of the Earth; a product of 'expansions and contractions' of the upper part of the crust during cooling. Having found that oceans provided an unexpected additional pull of gravity, Pratt considered it was the pull of the ocean away from the Himalayas that was assisting in limiting the plumbline's deflection. The discovery that gravitationally mountains were hollow, and that the oceans contained invisible mountains of mass, sustained one quarrel within the greater battle of geologists and physicists, fluidists against solidists. Airy's combination of fluidism with physics would remain lost for 20 years. From 1854 to 1858 the molten Earth flourished with the more or less unopposed blessing of a physicist. It was at this period that Green first had intimations of his molten globe. The door was only temporarily ajar; after 1860, for almost a century, it was never to open so wide again.

Clarence Dutton was born in 1841 and attended Yale University where he studied Mathematics and Physics. At the start of the Civil War he had joined the Ordnance Corps on a permanent officer's

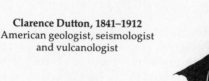

Clarence Dutton, 1841–1912
American geologist, seismologist
and vulcanologist

commission, and at the end of the fighting the desire for further adventure led him into the most exciting and challenging field of endeavour in the immediate post-war period – Geology.

Among Dutton's Washington contacts were Ulysses S. Grant and Major John Wesley Powell, the leader of the most successful and daring of the Western civilian survey teams. In 1875 Dutton obtained permission, by special act of Congress, to join the next Powell expedition to the Colorado Plateau. Dutton, Powell and Grove Karl Gilbert, the three intelligences of the party, were embarking on the greatest geological adventure in the history of the science; they were to turn the static principles of British physicists into dynamic processes, to see that a landscape was no less an evolutionary system as life itself. Above all they were to mediate the American West into the American consciousness. Of the three, Dutton had the natural artistic intimacy. His powers of exposition and the restraint of poetry were most tested by his encounter with the Grand Canyon; his expressive descriptions of the geomorphic forms of the Canyon that borrow a vocabulary from architecture and sculpture were published in 1881.[22] Fisher extensively quoted Dutton's observations on the uplift and volcanic activity of the Colorado Plateau as they provided confirmation for his own liquid earth. Dutton in return enthusiastically reviewed Fisher's work in the *American Journal of Science*.[23] In 1882 Dutton took time off from the survey in order to visit Hawaii, where he was shown around the volcanoes and encountered William Lowthian Green. These exchanges enforce the sense that the freethinkers of the Earth were a distinct group; in exile and disparate, yet united in viewing the Earth from a position distinct from that of Geology.

In 1886 the largest earthquake on mainland USA during the last quarter of the century devastated Charleston, South Carolina, and was felt along the entire eastern seaboard. As a demonstration of his role as a natural philosopher Dutton went to study the earthquake and wrote the official report on the disaster.[24] Further travels took him to the great volcanoes of Central America. Thus at the end of the 1880s Dutton had covered the whole gamut of Earth behaviour. He was acknowledged as the US's greatest expert on volcanoes, and after his 1904 textbook on Seismology had capped his report on the 1886 earthquake, the greatest influence in the study of Seismology, effectively bringing the science from Europe to America.

Dutton viewed the Earth as process; thus he broke away from the materialism, the rock collection, that was at the heart of Geology. His 1881 report on the Colorado Plateau presented a journey through the region as a journey through time. Erosion was balanced against vulcanism; the scenic evolution of a day's observation was also the evolution of the landscape. Dutton was not to be drawn into extrapolating from his observations into a blueprint for the total Earth. However, on 27 April 1889 he was offered a vacant half hour in a scheduled regular meeting of the Washington Philosophical Society and he was listed to speak on 'Some of the greater problems of physical geology'.[25] He had three in mind: 'What is the cause of volcanic action?', 'What is the cause of the elevation and subsidence of restricted areas of the earth's crust?' and 'What is the cause of folding and distortions and fracture of the strata?' He spoke out against the simple notion of a contracting Earth as an explanation of mountains; of the mountains of his experience where he had passed great years in the American West. All the evidence pointed to there being a plastic substratum; crustal rebound had already been demonstrated around the Baltic Coast after the supposed melting of a North European ice-cap. He cited the Greenland ice-cap as to how rocks might behave under pressure.

Seven years earlier in his review of Fisher's *Physics of the Earth's crust* he had hinted at an unpublished paper that introduced a new term, 'isostasy', for the condition of equilibrium of figure 'to which gravitation tends to reduce a planetary body'. Now in 1889 he explained how the word was to provide a raft to help a mind at sea. Isostasy offered explanations for local vertical movements of readjustment and for local topographical equilibrium. Thus, as the Colorado River carves the Grand Canyon, so the land will rise in compensation for the loss of weight. Isostasy was a principle by which other explanations could be tested.

Dutton's neologism had immediate currency. In 1888 Chamberlin had been mapping the shorelines around the Great Lakes that had formerly been weighed down beneath a massive ice sheet. He showed that the region was tilting as the north rose faster than the south, one day to send the lake water flowing into the Mississippi.[26] Also in 1888 Gilbert had been mapping the former shorelines of Lake Bonneville in Utah, a lake that during the Ice Age had been far more extensive. Once the water level had receded and the weight removed, these too he claimed had been deformed as the crust was warped during uplift.[27]

The following year Dutton ended his full-time geological furlough, returning to be a major in charge of the San Antonio arsenal, from whence he could mount forays to Mexican volcanoes.

The mountain builders

North American Geology had matured along very different lines from its British parent. As Dutton named 'isostasy' so it was Grove Karl Gilbert who had coined 'orogeny' to mean the overall creation of mountain ranges, and had recognised thereby the appropriate scale of abstraction necessary to encompass such vast, lengthy and unseen processes.[28] It was Grove Karl Gilbert who was working as an extra-terrestrial geologist even before Chamberlin, performing stellar observations from the base of the Grand Canyon while travelling with Powell and Dutton, and confessing that he was 'a little daft on the subject of the moon'.[28]

In 1891 Gilbert made detailed investigations of Coon Butte in

Arizona (now renamed Meteor Crater) that he suspected as being a product of some collision with an asteroid.[28] Through 1892 Gilbert performed a series of experiments in a variety of Western hotel rooms, firing projectiles into packing cases of mud until he had demonstrated to his own satisfaction that the form of the craters on the Moon were not, as had been assumed, the product of volcanic activity, but were produced by asteroid impacts – a discovery that he reported as a lecture in December 1892.[29]

Part of Gilbert's desire to explain the Moon's surface as a result of asteroids, as also Chamberlin's desire to make them the centrepiece of his cosmogony, was that they were a product of American science. Kirkwood, the champion of asteroids, had been at the University of Indiana. Away from the East Coast that was prone to European fashions and culture, this mid-West region was honing a raw and original America. The man who was to propose the American version of the moving continents, Frank Bursley Taylor, was born in 1860 at the very heart of the Cosmology Belt – Fort Wayne, Indiana. In an age populated by tyrants like Chamberlin, intellectual aesthetes like Dutton and great experimentalists in landscape like Gilbert, it is ironic that continental drift in a form recognisably that of the 20th century should have been proposed by a cranky and unprepossessing glacial geomorphologist.

Taylor's father was a prominent and successful patent lawyer with a great scientific curiosity. Frank was an only child, pampered and frail. In 1882 he entered Harvard as a special student where he took courses in Geology and Astronomy. In 1886 bad health forced him to leave without achieving a degree and in order to 'gain vigour' he spent the next 13 years travelling around the region of the Great Lakes with a personal physician. To give purpose to the carriage rides he mapped the raised shorelines, overflow channels and moraine hills of the most

Frank Bursley Taylor, 1860–1938
American glacial geomorphologist

recent ice sheet. His father paid for the physician, the travelling and, as the work matured, the publishing expenses. These obsessive recuperative studies gradually won for Taylor a certain respect; Gilbert even passed on some unfinished notes. Although an active member of various geological and scientific societies, Taylor existed for much of his life in isolation from the geological community, being supported by his father until 1900 and by his father's estate from 1916 until his death. This isolation gave him considerable freedom; both to speculate and to indulge in theory building away from the possibility of effective criticism. A decade of reconstitution and some inspiration gained from Chamberlin's report of the Greenland expedition of 1896 led Taylor in 1898 to write up his personal cosmology as a privately published 40-page pamphlet.[30] This theory owed its origins to his Harvard Astronomy courses and most significantly to Daniel Kirkwood's asteroids.

Taylor's Moon was a comet that had recently been captured by the Earth. As comets grazed the Sun they also became captured in the orbit of Mercury, and as every new comet was captured, each planet shuffled out one orbit towards Neptune. Like Kepler, Taylor believed that the planetary orbits were all determinate, fixed paths with no possibility of an intermediate course. The Sun gains a series of adopted children; each new arrival displacing the position of the previous youngest. In 1899 immediately after completing this pamphlet, the physician was displaced by a new arrival – a wife. Minetta Amelia Ketchum came from the site of his first geological investigations and the site of his first paper. As the orbit of Mercury was determinate, so his wife's orbit was that followed by the physician. She accompanied him on all his field trips, taking the reins of the horses and, later, the steering wheel of the automobile.

The arrival of the Moon to the Earth, as a wife to this earlier adopted son, had seriously perturbed the order of that planetary establishment, creating a strong tidal force and increasing the speed of rotation. These resultant forces had pulled the continents away from the Poles towards the Equator. This story of moving continents was, in the 1898 paper, little substantiated. The movement from the Poles seems most likely to have been contrived to explain the 'imaginary' former ice sheets with which Taylor's investigations had been totally concerned. Like Chamberlin before him, he had spent more than a decade investigating the footprints of an ice sheet that he had never seen. Perhaps the ice had spread out from the Poles like the icing on a spinning cake. Perhaps, and this was a germ of an idea to re-emerge in Europe, the great ice-maker Greenland had once been locked to its neighbouring continents.

The marriage served to celebrate the beginning of a phase of employment, as a special assistant in the Glacial Division of the US Geological survey under the supervision of Chamberlin. In 1903 his father convinced him to turn the cosmology into a book 'to put on record his son's great ideas.'[31] Taylor used the opportunity to detail the mathematics of the theory. This was heady stuff, opposing Newtonian mechanics because it did not provide a single determinate orbit for the moon. In this racetrack solar system there was no mention of his 1898 continental movements. He put these to one side because in the years since 1898 he had been avidly reading each volume of Suess's *The face of the Earth* as it was translated into English. In marriage Taylor had been frequenting his armchair where he self-confessedly did his theorising. At the back of his mind there were always the ice sheets; he never considered that the rocks of Earth were implicated in his theories; the continents moved like ice sheets, over the surface of an inner planet that was more rigid than the hardest steel.

Taylor's cosmology, like Chamberlin's, was a response and a challenge to the doomed Victorian 'Earth' falling into degeneration and decline. The more he read Suess, describing the remarkable youth of the Earth's mountains and the impression of great outflowing of the continents, 'The whole southern border of Eurasia advances in a series of great folds towards Indo-Africa', the more Taylor saw mountains to be no more than moraines, running east–west just as the moraines struck out across the only landscape he knew – that of the mid-West. Along with many other readers anxiously awaiting the next volume of the master work, there was an expectation that the Austrian was teasing with clues and would eventually provide some grander scale of Explanation. For Suess himself lived in the foothills of the Alps, a mountain range that during the 19th century had begun to reveal a most unlikely geological story.

In 1841 a Swiss geologist, Escher van der Linth, had been mapping in the Glarus district when he discovered a sequence of geological strata, which persisted from the Wallen See to the Rhine Valley, in which the older was on top of the younger; highly coloured purple-red sandstones of the Permian age, above pale sediments of the Mesozoic and Tertiary ages. The configuration was indisputable; his explanation presented in a short lecture was that the rocks had become involved in two giant overfolds, one that had advanced from the south, a second from the north. To fulfil this contortion at least 15 km of horizontal movement was required. This was unthinkable: 'No one would believe me if I published my sections', claimed Escher 'they would put me in an asylum'.[32] The information stayed hidden, Escher's sanity unquestioned, until his sections were published by another Alpine geologist in 1875.[33] Yet worse lunacy was to follow.

Marcel-Alexandre Bertrand, born in 1847, was a French geologist who, through his mathematician father and training at the École Polytechnique and the École des Mines in Paris, followed in the footsteps of Elie de Beaumont. Bertrand mixed geological fieldwork with a desire for a grander overview of the globe and its distinct phases of mountain building, reviving the theories of William Lowthian Green (supported by the French crystallographer Michel Levy) for the tetrahedral Earth. In 1884 Bertrand proposed that Escher's two separate folds were in fact one single great overthrust; an

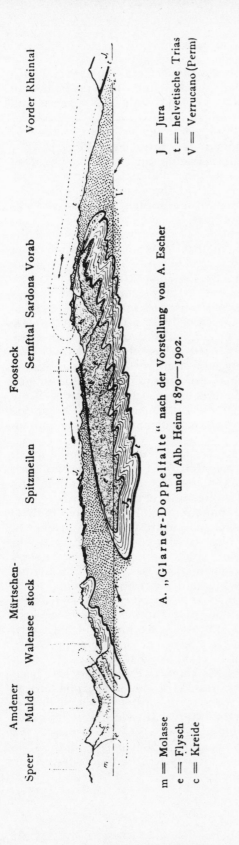

Sper Amdener Müртschen-
Mulde Walensee stock Spitzmeilen Foostock
Sernfttal Sardona Vorab Vorder Rheintal

A. „Glarner-Doppelfalte" nach der Vorstellung von A. Escher
und Alb. Heim 1870—1902.

m = Molasse
e = Flysch
c = Kreide

J = Jura
t = helvetische Trias
V = Verrucano (Perm)

B. „Glarner-Deckfalten" nach der Vorstellung von M. Bertrand 1883 und E. Sueß 1892, angenommen von Alb. Heim 1903.
Die Glarner-Deckfalten in schematischem Profil von N nach S nach älterer und neuerer Auffassung.

Escher van der Linth's 1841 cross section of the Glarus district, originally indicated as two folds but reinterpreted as one major nappe by Bertrand
in 1883

explanation that required horizontal movements of around 100 km.[34] However, other examples of these *grandes nappes* (from the French for table-cloth) were soon found elsewhere in the Alps and Pyrenees and such gigantic overthrusts were beginning to turn up in the older mountain chains.[35, 36] Nappes became 'officially recognised' at the 1903 International Geological Congress in Suess's own Vienna. They provided incontrovertible evidence that mountain building involved massive horizontal crustal shortening. While such shortening was compatible with a contracting Earth, the manner in which the movement was concentrated on narrow mountain zones was stretching the theory towards its elastic limit.

It was left to Suess to extract from all the new observations of the mountain ranges, explanations and theories for their origin and he set out to process and interpret geological details from across the globe. He had described in his original *Die Enstehung der Alpen* of 1875 how mountain ranges were asymmetric; with the formation of a major fracture the crust had become overthrust through continuing global contraction. He saw that the trends of folds around the Alps were influenced by the buttressing of rigid blocks of continent, such as the Bohemian Massif, located to the north. These rigid blocks had once themselves been mountain ranges but were now 'stabilised'. After a superlative effort to take command of the rapidly expanding information, in a move that could never be repeated, Suess began his synthesis of the world's geology. *Das Antlitz der Erde* opens with the Alps as the archetype and travels out to include all other mountain ranges.

Suess then proceeds to theorise. The horizontal movements implicit in mountain ranges confirmed a contracting Earth. On such an Earth the oceans were collapsed continents. Having failed to find any modern 'geosynclines' located in the midst of continents where they have today been turned into mountain ranges, he preferred to deny the existence of such phenomena past or present. In the defence of his contracting Earth, Suess also prohibited any primary uplift, considering that all explanation could be sought in downward displacements, horizontal movements and the changes in relative sea level. Thus his unshakeable faith in the contracting Earth determined a number of his conclusions. However, Suess's greatness is as a syncretist: within his epic compendium of world geology and, most significantly, within his ready acceptance of horizontal deformation and the larger scale with which he considered mountain ranges, he prepared a body of information that could be given an alternative interpretation once his implicit contraction had been challenged. On a number of occasions his descriptions of phenomena are inconsistent with a simple contracting Earth; for example in 1892 he wrote that the East African and Rhine rift valleys were tensional phenomena.[37] In his final volume he toyed with, and dismissed, the possibility that 'tidal forces and rotational phenomena', in the style of the 19th-century fluidists, could somehow be implicated in the formation of mountains; but at the age of 78 it was too late for any intimations of doubt to turn into apostasy.

Taylor waited until the final volume of Suess's great work had appeared and then, in an address delivered at Baltimore on 29 December 1908 (published in 1910[38]), presented his theory of continental movements. The Equator-ward creeping of the continents had impinged on the great fixed bulk of Africa and India, whereby the greatest mountains had been uplifted. While Suess considered that there had been at most some kind of outflowing at the margins of the continents with no displacement of the continental cores, Taylor, taking all his primary data from Suess, considered the whole continents to have moved. More importantly, as mountains were moraines, so some of the oceans were crevasses. The similarity of forms between the two sides of the Atlantic, and the submerged mid-Atlantic chain of mountains already known to echo the continental coastlines, were taken by him as indicating some past contiguity. A more perfect fit could be found between the West Coast of Greenland and the islands of north-east Canada – the two had moved apart by several hundred kilometres. Greenland was shown to be part of America. At the north-west corner of Greenland there lies one of those rare locations on Earth where continental drift seemed almost visible. In 1910 Robert Peary was to remark of the extraordinary straight-sided Robeson Channel

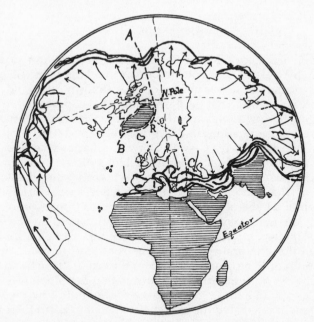

Taylor's continents spread towards Africa and the Equator and away from Greenland (from American Association of Petroleum Geologists 1926 symposium on continental drift)

'whether it is a Titanic cleft formed by the breaking off of Greenland from Grant Land, is a question still unanswered by geologists.'[39]

The detail present in these North Atlantic reconstructions makes Taylor the first to outline a continental drift with some of its 20th-century character. However, the analogy with the ice sheets was too strong, and his own training as a glacial geologist could not hope to provide any further contribution to the theory. Thus Taylor, creating a cosmology in isolation, was condemned endlessly to repeat his 1908 lecture.

Taylor preserved a geologist's conception of the Earth, in which the internal constituents were structurally inert and impotent in affecting the surface behaviour that remained the object of geological investigation. That his planet was 'as hard as steel' conveniently removed it from all further consideration. The cause of the continental movements had to come from outside. Taylor's cosmology had already provided such a mechanism, but now he removed from it all the history of the Sun's children and restricted it to the Earth gaining a lunar wife. The Moon had arrived in the Earth's later life, during the Cretaceous geological period, and had put the Earth into a faster spin that had pulled out its midriff. The continents slid from the Poles in response to increased rotation. Taylor suffered the dubious fate awaiting almost all of those who proposed whole-Earth theories. Neither mocked nor humiliated, he was ignored.

Sister Moon

The history of Taylor's Moon flew in direct opposition to a story proposed in England. The Pacific Ocean, the largest of the oceans, awaited explanation. It was almost circular in form, surrounded by a ring of fiery volcanoes. That the Earth should have spawned the Moon, a parthogenetic birth, was an idea that even its 19th-century proponents recognised had mythic origins. In the mid-19th century the story became blessed by George Darwin, Charles's second son, a physicist performing research under Kelvin's direction. Darwin had not set out to prove any Moon birth but his investigations had become focused on the Moon's effect on a molten Earth purely as an application of a general problem. Darwin asked himself under what conditions the Earth day and the lunar month would be of the same length. Of the two results, 55 days and 5 hours and 36 minutes, he considered that the first result must be some future state while the second result could be of significance for the origin of the Earth–Moon system. The 5½ hour month-day required that the Moon was within 10000 miles (16000 km) of the Earth. At 5 hours the two would be united. Darwin calculated the energy dissipated in Earth tides within a molten planet, allowing the Moon to spiral out to its present orbit, and found that at some time, more than 54 million years before the Earth became solid, the Moon was born out of the Earth.[40]

Darwin's story of Moon delivery soon became very popular. His age estimate, itself a minimum, was in respect of a molten Earth, but in the way of all numbers became paraded alongside those timescales obtained by Kelvin, which found from heat flow the age of the geological Earth, the age since solidification. Darwin attempted to adjust his history by claiming that the Earth was still sufficiently viscous to possess some unseen Earth tides. Adjustment and counter-adjustment followed, yet the power of the story of the birth of the Moon was sufficient to sustain it independent of all the necessary detail. Darwin's original mother Earth was entirely molten, but Fisher, writing in *Nature* in 1882,[41] preferred to allow the birth to have geological significance and suggested that the scar would be unlikely to heal so completely, in particular if it took place relatively late in the formation of the Earth. If the Moon was made predominantly out of terrestrial crust that would explain its low density. The great crustal rug had been full of water as it was flung off and wrapped into a tidy ball; water that provided the widespread 'vulcanism' that could be seen to have peppered the Moon's face.

Fisher's scar was the unique and circular Pacific Ocean, for the Earth had once been enveloped in a crustal rind and the continents and islands, which now made up only one-quarter of the Earth's surface, were all that had failed to join the moonride. In the trauma of birth, as Fisher described it, the remaining crust had become broken into islands, and on the far side of the globe to the Pacific, the stretch marks were still visible where the Atlantic Ocean was a 'great rent', as demonstrated by 'the crude parallelism that exists between the contours of America and the Old World'. Around the Pacific, the pattern of fragmented crust was at its most complex where the currents at the margin of the scar persist to this day. Fisher expresses no suggestion that the recognition of the 'crude parallelism' of the Atlantic is his own; rather that it is self-evident and perhaps awaiting explanation.

In 1897 Darwin crossed the Atlantic to give a series of influential lectures at important East Coast societies and universities. This visit provided a rare contact between American and English cultures; for many in his audience the Moon birth was completely novel. Among these was an astronomer, William Henry Pickering. He was born in Boston in 1858 and had joined Harvard University in 1887, where his brother, the great astronomer Edward Charles Pickering, was already director of the Observatory. William was in danger of being perma-nently eclipsed by his famous elder brother and he was happy to be perturbed by Darwin's lecture to make an almost unprecedented step for an astronomer, of turning his attention to the Earth. Pickering toured the craters of Hawaii in 1905 and those of the Azores in 1907 in order to make morphological comparisons with those to be found on the Moon. In 1907 he published in Chamberlin's own *Journal of Geology* 'The place of the origin of the Moon – the volcanic problem',[42] in which he began by restating Darwin's hypothesis but then went on to disavow any past liquid Earth; possibly to evade the censorious editor's scissors. Figure Two in Pickering's paper is a plot of depth-heights for the Earth's outer surface, taken from Gilbert.[43] Pickering identifies the two average elevations (the ocean-floor and the land close to sea-level) on this diagram, and also the true edge of the continental plateau that he places at 1000 ft (300 m) below sea level. He then repeats Fisher in claiming that the Moon emerged out of the Pacific dragging the other matching continents apart, and calculates that a crust 36 miles (60 km) thick extending over all the modern oceans of the Earth would make up the volume of the Moon. The Moon's maria are matched with the arcs of volcanoes, as those around the China Sea. Observations of the pull of gravity made on the edge of the Hawaiian volcanoes suggested that the lower half of these mountains has a lunar density. In the middle of the Pacific Hawaii was the eroded umbilical: the omphalos. No need for extra-terrestrial journeying when here on Earth, there existed the rocks of the Hawaiian Moon.

Without placing a date on this great process of fissioning, Pickering renovated the Moon birth story. The paper was so well received that it was reprinted in the *Scottish Journal of Geography*, and translated into German to appear in the *Beiträge Geophysik Leipzig*, 1909. As Pepper had shown with Snider-Pellegrini's Origin of the Atlantic, whether such theories were catastrophist or gradualist was almost incidental; the story could survive any distortions of time.

The origin of the continents and oceans

We have come to the end of the creative story in North America. Chamberlin's planetissimal theory had so totally integrated Geology with Astronomy that it could now effectively throttle other theories of the Earth at birth. Some theories had, however, predated Chamberlin's; the healthiest of such rivals had been proposed by a geologist Bailey Willis in 1893[44] before the planetissimal theory was even a twinkle in its father's eye. Willis had been inspired by Dutton's account of isostasy to contemplate the significance of this crustal balancing and had reasoned that as continental rocks were eroded and washed away into the oceans there must be a corresponding under-current of material in the substratum displaced by the heavier oceans, moving under the lighter continents and pushing up mountains. The encroaching oceans were the antithesis of Suess's outward bulging continents.

Across the Atlantic France possessed a more liberal and dynamic tradition of Geology, in particular through the studies of the Alps. The legacy of Elie de Beaumont sustained a search for higher order in the configuration of the Earth. France had also had its 'William Lowthian Green'. M. R. Mantovani lived on the distant French volcanic island of Réunion, set in the Indian Ocean, where he had pondered on the shapes of the southern continents. In December 1889 Mantovani

W. H. Pickering's opening of the Atlantic following the loss of the Moon from the Pacific Ocean (from *Journal of Geology* 1907)

presented a paper in Réunion,[45] in which he explained that the continents grouped around the South Pole had opened out like a flick of a lady's fan. It was not until 1909 in a newly formed popular Parisian magazine (*Je m'instruis*) that his adapted theories could hope to find a wider audience. To improve the image of fanning Mantovani had taken some (prescient) liberties with the extremely ill defined Antarctica; dividing the southern continent into two unequal pieces in order to turn both Africa and Australia into replicas of South America, with southern tails as sharp as that of Tierra del Fuego. Mantovani's continental fan had opened through Earth expansion.

The one country with a strong scientific tradition that was still relatively free from any orthodox opinion on the subject of the whole-Earth was Germany. During the great German revival of the late 19th century the Government had been encouraging scientific travellers.

Mantovani's 1889 rearranged continents around the South Pole – prior to Earth expansion

They had succeeded in collating the information on fossils and on past climates that had contributed so much to Suess's great synthesis. The coal deposits of North Greenland were matched by evidence for a great ice sheet in tropical Brazil. The past world seemed topsy-turvy. It was fitting that it was a researcher in the great German science of Meteorology who was to propose the new theory of continental movements for the difference between Alfred Wegener and all his predecessors was not that his theory was markedly more scientific than their's, but simply that within the study of past climates, he had the possibility of undertaking active research in order to attempt to obtain verification in the manner of a scientist.

On 6 January 1912 Wegener, the 31-year-old meteorologist, gave a talk to the Geological Society in Frankfurt, about a series of ideas developed over the previous two years. The origin of these ideas he claimed had come through observations of the parallel coastlines of the Atlantic. There are many sources that could have triggered some association, reading Peary, or Pickering's translated article, finding confirmation within Suess. The first talk was cautiously titled, with respect to his geological audience, 'The building of the gross form of the Earth's crust'.

The structure of the argument that he presented in this first lecture was to remain largely unaltered through all his subsequent publications. Following Pickering he matched the Atlantic at the continental shelf break, and also gave great emphasis to the distribution of depth-heights which revealed, through the mediation of isostasy, that the crust of the continents and that beneath the oceans was built out of a different material; the continents being made of Suess's sal (later altered to sial) that floated in the substratum of sima exposed in the ocean floor. The present outline of the continents to either side of the Atlantic was a result of the disintegration of an original super-continent during the Mesozoic geological era. Evidence for former contiguity of the two sides of the Atlantic could be found in the position of the older mountain chains, from the matching of former ice ages, the moraines of the last Ice Age, and from the evidence of the fossils. The palaeoclimatic reconstructions also indicated that the continents had moved *en masse* in relation to the poles. These continents were motivated by forces associated with the Earth's rotation; the same forces that were responsible for the movements of the atmosphere.. There was the Eötvös force that drove the continents away from the Poles, and the tidal force from the Sun and Moon that caused the continents (as proposed by Green) to lag behind the revolving Earth and so shift to the west. The only proof of the continental displacement, Wegener claimed, would come from astronomical observations of longitude made between the continents; observations that for Greenland already suggested some demonstration of movement.

The response was hostile, the geologists were not to be seduced with such effortless globalism. Four days after this first talk, Wegener found an audience that was to prove more sympathetic to his moving continents. He presented the story under the more audacious title 'The horizontal displacement of the continents' to a general scientific public at the Society for the Advancement of Natural Science in Marburg. To a geologist, Wegener's theories were about the Earth and yet were totally ungeological. To a non-geologist, unaware of the culture of that science, Wegener appeared to be presenting ideas greater and more daring than those of geologists. Just as his argument with its mixture of vision and confusion was to remain structurally unaltered to his death, so the reception for his theory was effectively cast at these first two meetings. Soon after, with his talk published,[46, 47] Wegener left on his third hazardous expedition to Greenland with the Danish explorer J. P. Koch. He returned in 1913, and found, owing to the illness of Koch, that he would have to write up all the copious expedition notes on meteorology. For three years, from the birth of the moving continents, he had been in close communication with the 'grand old man' of German meteorology, Professor W. Köppen of Hamburg, and in 1913, in a strategic union, married Köppen's daughter Else, a woman who shared Wegener's own culture of scientific curiosity.

Dutton had died at Englewood, New Jersey, two days before Wegener gave his first lecture on the floating mobile continents, that had drawn so heavily on the American's isostasy. Fisher had passed away in 1914, three years short of his century. Taylor was keeping his head down, still an assistant in the glacial division of the US Geological Survey. And the famous geologist William Henry Pickering, born in 1858, 'on July 9th 1912 while heroically leading a party at Cadeby Colliery lost his life through the gas exploding a second time'.[48] History plays strange tricks; William Henry Pickering, born in 1858, not the leading geologist and President and founder of the Indian Mining and Geological Club, but the American astronomer, had gone into hiding. In reading the obituary the geological public could be deceived into believing that the author of the theory of the origin of the Moon and of continental separation was no more. The astronomer had merely slipped out of the mainstream, in 1911 moving to the tropics of Jamaica to run Harvard's new observatory. On 29 July 1914, before he could return to the moving continents, Wegener found himself caught up in an unforeseen tragedy. By the autumn of 1914 he had enlisted in the Kaiser's army as a lieutenant in the third division of Queen Elizabeth's grenadier guards and was off to the treacherous Belgian front.

4

Coming apart at the seams

Wegener's drift

Alfred Wegener, born in Berlin in 1880, was the youngest son of a protestant minister. As a schoolboy he became fascinated by the tales of the exploration of Greenland, *the* great destination of the 1890s, providing the possibility of a land-based route all the way to the North Pole; and even after Greenland had been shown to be an island giving way before the Arctic Ocean; yet it was Greenland that had become imprinted on him as his own personal goal. Throughout his life the bond he sustained with his three-year-older brother Kurt was that of a twin; their paths were destined perpetually to intertwine. Perhaps because his brother had already chosen to become a specialist in Meteorology, Alfred moved into Astronomy commencing a doctorate achieved under the supervision of Wilhelm Forster at Berlin and completed in 1905.[1] The scope afforded by Astronomy could not have prepared him for the nature of his toil: interminable arithmetical recalculation of medieval tables of planetary motion. He achieved the doctorate, published an article on astronomical history[2] and thereby ended his career as an astronomer. His chief memento from this the final stage in his education was a distaste of all things mathematical: in a letter sent by Alfred to his future father-in-law he states 'I hold the crass and probably exaggerated point of view that such mathematical treatises as I cannot understand are wrong or do not make sense'.[3] The mathematics became his justification for quitting Astronomy but this is of the form of an alibi; the arithmetic of his doctorate had been repetitive rather than taxing.

The attempt to diverge from his brother had failed; Alfred had been attending Meteorology seminars while writing his thesis and by 1905 had even published his first meteorological paper on the scientific uses of kite-flying in thunderstorms.[4] Meteorology was enjoying a considerable vogue. The stratosphere had only recently been discovered, and its influence on the weather remained to be found.

Alfred Lothar Wegener, 1880–1930
German meteorologist and
geophysicist

Wegener took up an assistantship at the Aeronautical Observatory in Lindenburg to research into the movements of the upper atmosphere.

To observe these remote cloud kingdoms required kites and balloons. High altitude ballooning therefore became legitimised as integral to the science, and also provided one other satisfaction: fulfilment of his cryophilia. Living in North Germany, how can one more readily reach polar conditions than by ascending in a balloon? If one incident were to be isolated as providing an insight, a shaft of light into the man, then it would be the 'Arctic' flight that he undertook with his brother, high above the German countryside, setting off on a spring morning in 1906. The 1200 m³ balloon was launched from the Aerological Institute at Berlin. The Wegener brothers first drifted north towards Denmark, out across the open sea of the Kattegat, before the shifting sluggish winds drew them to the south, to a landing 52 hours later near Frankfurt. The world record for flight endurance was beaten by 17 hours. By day they flew at 5000 m; by night they would descend to avoid the scourge of frosting, settling in the Arctic, −16°C at 3700 m. The transition from night to day was the transition from Astronomy to Meteorology. For the first time Wegener could look down and see the Earth, not as a geologist sees it, but as a fabric as extensive as the sky; a second sky in opposition to that of the clouds. The Earth as sky; this was to be Wegener's endeavour.

The distant love-affair with Greenland was soon to reach a resolution. In 1906 he was invited, as the sole German, to join a Danish expedition to north-east Greenland, led by Mylius Erichsen, that was to involve spending two winters at 77°N. Greenland had gained great significance within the new Meteorology, for as recently as 1890 it had been thought that European weather travelled from North America intact, though perhaps becoming a little tangled. Details of the storms in New York could serve to provide five days warning for sailors to batten down the hatches in the German Ocean. The situation was now recognised as being far more complicated, and the demon of disorder was Greenland; a vast continent of ice moored within striking distance of the shipping lanes of the North Atlantic. But Greenland was also the white, empty continent, cold, clean, a world of ice in which Wegener had the opportunity to study everything in the world: Astronomy, Meteorology, Glaciology – everything except life. The cryophile avoids the sensual tropics for the sake of his soul.

In the spring of 1908, purified, Wegener returned to Germany where he settled in Marburg and began to lecture. The next 3½ years before the compulsion to return to Greenland overtook him once again were the great creative period of his life. His biographer Benndorf has written on this period as 'the spring that follows the cruel arctic winter, when out of the barren tundra fresh grass and flowers burst out that have been nurtured unseen in the soil'.[5] 'Wegener's imagination overflowed like the brook that is swollen by the melting snows', etc. Among these blooms there appeared a little primrose known as continental drift. His main output was, however, meteorological and in 1911 he had completed his first book *The thermodynamics of the atmosphere*. This had been preceded by many papers; but where in 1909 he wrote 'On the origin of cumulus mammatus' by 1911 it was 'On the origin of the continents' (Die Enstehung der Kontinente).

Soon after Wegener had reached his majority, in time with the beginning of a new century, new clues had been gained towards solving the mystery of the Earth. The most sensational of these were the ramifications of radioactivity, which first fogged the photographic plates of Henri Becquerel in 1896. Such strange emanations had no significance for geologists until 16 March 1903 when Pierre Curie announced that radium salts were continuously releasing heat.[6] Within a few months, radioactive decay was summoned to explain the long-lasting energy of the sun;[7] and in the spring of 1904, with the discovery of significant quantities of radioactive elements in ordinary soil, the first suppositions of how such heat could contribute to the Earth's own internal temperature, also became fortified.[8]

Kelvin was intrigued by radioactivity, but never publicly recanted his impositions on the age of the Earth or the significance of any additional source of heating, preferring to believe that radiation was the release of some stored energy, perhaps obtained out of the aether. Yet to those such as Dutton and Fisher, who had spent their lives fighting against the notion of a young, dead Earth, radioactivity vindicated those intuitive opinions; the Earth of the new century had its own heat source and was alive. Even George Darwin welcomed the opportunity offered to extend his own calculations. For the discovery was not just some addition to the knowledge of the Earth; its significance lay in the sense of liberation that it gave to those who had suffered under Kelvin's tyranny. It would take 20 years before physicists could even begin to re-establish their supremacy over geologists in defining the workings of the globe, and meantime, in their new-found freedom, geologists could theorise from their own observations, unfettered by the chains of a mathematical argument.

Another vindication of the geologists' beliefs came from experiments carried out by physicists at McGill, in Canada. In 1902,[9] after many attempts, they succeeded in causing a block of marble imbedded in a cast-iron sheath to bulge and bend without breaking. For the past 50 years geologists had seen rocks squeezed, pulled apart, folded and twisted. The triumvirate of the Powell Survey, Powell, Dutton and Gilbert, had all independently written of their beliefs that rocks at depth in the crust were, like glacier ice, solid but plastic. Now for the first time it was possible to reach the lower crust in the laboratory.

The bulging marble cylinder had little impact on the physicists because the hidden interior of the Earth had suddenly begun to be made visible. Seismology had provided a 'telescope' with which the interior of the Earth could be inspected. This, the most important bridge on the Astronomy – Earth Physics frontier, came in a brilliant piece of inductive science posing as a deductive argument, presented by Richard Dixon Oldham, a British geologist and seismologist who was for a long period Superintendent of the Geological Survey of India. His paper, presented in London in 1906,[10] begins with the intention to tread the path of Jules Verne, on a mental journey to the centre of the Earth: 'Of all regions of the earth, none invites speculation more than that which lies beneath our feet, and in none is speculation more dangerous'.

Three different kinds and speeds of waves arrived at a seismic instrument from a distant earthquake. The slowest of these was identified as a surface wave, and therefore of no interest to a study of the Earth's interior. The two remaining waves follow paths through the deep earth and therefore a study of the time taken for them to arrive at recording stations situated at increasing distances from the earthquake, can provide a crude picture of the distribution of velocities and hence densities of materials within the Earth. The first wave to arrive, Oldham identifies as a wave of compression; the second as a wave of shear. Close to the earthquake these waves are lost within an incoherent signal of disturbance. Oldham proposes that this demonstrates that the simpler properties of the substratum do not occur until depths of 'not more than a score of miles'. In 1909 a Yugoslav seismologist, Andreiji Mohorovičić, while studying the records of a local Croatian earthquake, was able to pin down this boundary between the crust and substratum (or mantle) and thereby have his name labelled against it: the Mohorovičić Discontinuity.[11] But Oldham is pursuing a larger, more elusive quarry. At greater distances from the earthquake source the average speed of the waves continues to increase, suggesting that the paths of the waves are passing through denser rocks with depth. And then at an angle of 130° from the earthquake the second wave cuts out. Something within the

Earth has cast a seismic shadow. At the heart of the Earth there must exist a hidden sphere of material entirely different from its surrounds. Subsequently it was realised that the inner sphere is opaque to shear, as only a liquid can be.

Through his new 'telescope' Oldham had demonstrated the most important geological boundary of the planet: where the mantle meets the core. He had also initiated a whole new science; a science that was to discover further boundaries and that dealt for the first time in a fully three-dimensional abstract Earth. The discovery that shear waves could be transmitted through the bulk of the planet proved the final blow to any surviving fluidists who persisted in believing that the Earth was a thin-skinned bladder of magma. Shear waves could not pass through liquids. Apologists could claim that liquids under high pressures changed their behaviour, but this was little different from calling a liquid a solid. Fisher[12] in 1909 came up with one of the most ingenious defences detailing how the second waves were not shear waves at all, but compression waves slowed in their passage through the bubbles of gas, dissolved in the magma; bubbles that were 'known to exist' from the absence of earth tides. This was out of time. The more the seismological investigations became published, and the more credence they were given as providing the first true knowledge of the Earth's interior, the more solid and rigid the Earth seemed to become.

The last of the great innovations of the new century was not a piece of scientific work, but a catastrophe that befell the city of San Francisco, on 18 April 1906, two weeks after Wegener had begun his record-breaking balloon flight. Not since Krakatoa had a disaster become so famous; happening on, even slicing through, the doorstep of Pacific America. Of all earthquakes of the past two centuries none has been so imminent, accessible and self-explanatory. To a community of geologists insistent on a contracting Earth, this earthquake was a major assault. At a time when the debate as to the origin of earthquakes was tilted in favour of some massive subterranean explosion, a journey into the coastal villages of Marin and Sonoma Counties to the north of the Golden Gate revealed that the event was associated with horizontal movement along a vertical fault. In the absence of human construction such movement would be lost in the randomness of the landscape, so it was exactly where human activity (and therefore human observation) was concentrated that the movement was apparent. There was no possibility of argument about it: simple paling fences, banks, hedges and roads had all been offset. Horizontal displacement of 6 m was almost enough to set the bandwagon of continental drift in motion.

Such a disturbance required proper authoritative investigation, and as there was no science that naturally laid claim to the phenomenon, the Governor of California was forced to create some middle ground through nominating four geologists and four physicist-astronomers to serve on such a committee. Earthquakes lay midway between the rocks and the stars. Only Grove Karl Gilbert, who had experienced the shock from across the bay in Berkeley, had previous relevant experience, having studied and mapped recently active faults in the Eastern Sierra. The committee performed an admirable job in celebrating the event in two huge volumes, copiously illustrated with maps and photographs covering all aspects from geology to physics.[13] No report on any earthquake up to that time was so physically massive. The San Andreas Fault remains the most famous such feature on the planet. The sole omission from this effort was historical context: the very reason why the earthquake persists with its hold over the American imagination. (Details of a major Bay earthquake in 1868 had been suppressed to avoid perturbing potential immigrants.) Without history, each Californian dawn is a thrilling and unexpected sensation. Without history 6 m of horizontal movement was of no fundamental geological significance: a maverick, a fluke. As the English seismologist Charles Davison wrote of the coverage: 'If the Californian earthquake were the only shock known to mankind, the attention paid to it could hardly have been more exclusive.'[14]

The great debate

Reserve lieutenant Alfred Wegener was despatched amidst the massed German armies, to break through Belgium to Calais. In only four weeks, the armies gained 100000 casualties, many of whom were recent conscripts cut down by battle-trained British troops in the hedgerows west of Langemarck. Wegener was shot through the arm, and then two weeks after his return to active service, again, more seriously through the neck. This second, fortunate, bullet effectively got him out of the war. He returned to Germany on sick leave early in 1915, and after a period of convalescence returned to fight for Germany from the comparative safety of the front line field weather service, operating twice on the Western front and once in Bulgaria and Livonia.

The experience of war had confirmed in Wegener an already incipient pacifism. Through Köppen he came more and more into contact with the group of scientists who were affiliated to the *Monistenbund*. Under its leader, the chemist Wilhelm Ostwald, the monists believed in the 'energetic imperative' whereby society should be made happy and efficient through science. Within this *weltenschauung* Wegener's trips to Greenland were to seem as pilgrimages of absolution.

With the world around him gone mad, with German nationalism and militarism in full flood, Wegener employed the unreal sanctity of his wartime convalescence to turn the story of the moving continents into a book, published the same year, 1915. The 94-page work arrived into a nation at the high point of its destiny, yet if one was to map the history of this period only from Wegener's published works, they represent a surreal response to the world around.

His Meteorology was moving towards the extra-terrestrial; the marriage between Astronomy and Meteorology. In 1915 he wrote on the changing colours of meteors and in 1916, after observing a particularly brilliant meteor fall, he became interested in the journey of this cosmic debris. The fire in the night sky was also the arc of the tracers and the bright startling flash of an explosion. Over the next few years, in imitation of Grove Karl Gilbert, he experimented with projectiles fired into different materials to observe the form of the craters. This was a science for wartime, the fields of the Marne and the Somme had been wasted into a lunar landscape. As late as 1920, Wegener was producing more on craters of the Moon than on continents of the Earth. Like Gilbert he ascribed these craters to external bombardment.

At the end of the war, with Germany in chaos, close to revolution, Wegener was in a detached, almost playful; mood, writing on the hair-ice to be found in frozen rotten wood, and on waterspouts and whirlwinds. However, in 1919 he finally found secure employment, inheriting the directorship of the Meteorological Research Department of the German Marine Observatory at Hamburg. Wegener's father-in-law had been departmental head for 44 years, the whole history of the science of Meteorology. Wegener was his natural heir: from the very first Köppen had been an ardent supporter of his son-in-law's theories. For Köppen's greatest initiative had been to see how plants were sensitive indicators of the long-term weather, thereby allowing the synthesis of all those parameters of wind, sun, rain, snow and average and extreme temperatures into a single broad concept: climate. Together they had searched for geological evidence of past vegetation and past landforms, from deserts to glacial tills, to vindicate drift. In collaboration they had developed palaeoclimatology into a science – a science that provided direct support for the moving continents.

From the beginning, Wegener's theory provoked controversy. Was the Atlantic Ocean a widening smile or a frozen gaping yawn? Neither Taylor nor Pickering had found their audience, yet Wegener gained an immediate sympathetic response from those researchers in past or present Zoogeography whose results he had employed in his reconstructions. Any enthusiasm that they generated induced a counter-attack of increasing ferocity. Wegener's first audience of geologists continued to provide the opposition. Diener in 1915 wrote 'at first sight the hypothesis appears to have much of value to it, but on close analysis it turns out to be but a playing with actual possibilities'.[15] It was easier for Jaworski[16] in 1922 to name those geologists who offered Wegener unqualified written support (four of them, but two of these Dutch) rather than attempt to define the depth of the ranks of critics.

In 1917 Wegener received unsolicited corroboration for his theory with the publication of a survey among 20 palaeontologists of their requirements for land bridges between the continents at different geological periods.[17] Wegener took the information from this opinion poll and showed how much more economically drift could provide an explanation. The special affection he had for Greenland ensured for this subcontinent, as with Taylor, a pivotal position in his theory. While Wegener was a meteorologist, he showed no interest in observing a tropical typhoon, only a polar blizzard. Wegener continued to gather longitude determinations made from lunar observations on the west coast of Greenland in order to attempt to show that that continent, above all others, was engaged in rapid movement – a flight from Europe of around 10 m per year.

When in 1919 Wegener published these new observations there was a significant linguistic shift from the cumbersome *verschiebungen* (displacement), to the lighter active form in which the theory was to become popularised. [18] The *'Trift'* in *'Nordgrönlands Trift nach Westen'* was a word reserved for floating timber (as employed by Pepper) or for an ocean current. It implied a gentleness of movement, an effortless subtlety as possessed by wisps of smoke or clouds; nothing that was connected with the bulky, clumsy or resistant.

The new phase of employment was making Wegener bold. In 1919 he brought out a revised, second edition of *Die Enstehung der Kontinente und Ozeane*, extending it to 135 pages, and thus brought his theories afresh into the post-war German intellectual maelstrom. His map of a closed Atlantic had now become transferred onto a globe. India expanded, from an insubstantial triangle to a giant elongate peninsula thrusting all the way from Asia into the hollow between Australia and Africa, hard up against Antarctica. This single global continent was a united landmass; a Pangaea (as a Greek compound), a term that was taken by his readers to be the actual name of the continent. Whereas the comments on Wegener had previously been a wartime whisper, now they were to become a clamour.

While he could remain unconcerned by his reception from geologists, Wegener depended on the support of the geophysicists; for Wegener himself, as a meteorologist, was also part of the German school of Geophysics. This was therefore judgement by his peers. In 1921 Wegener could state that he knew of no geophysicist who opposed the drift hypothesis; but there were, as he also knew, many who were sceptical. During 1921 he was invited to write the summary to a series of lengthy, sober, critical articles produced by two young Austrian structural geologists who were strongly antipathetic: a geophysicist, Walter Schweydar, who was prepared to offer limited support, and Professor Albrecht Penck, a distinguished Austrian geographer, with a global view ('his eyes were forever on the far horizon'), who extended almost total endorsement. [19] The problem as ever was the speed Wegener demanded for the westward drift of Greenland. Jaworski in 1922 summoned cruel disbelief to Wegener's most recent claim for 1190 m of westward drift in 37 years, sardonically reflecting that in the Alps such rates of movement would certainly be 'impressive'.

Wegener's reputation had also broken out of Germany, first into Scandinavia and Holland and subsequently into France and England.

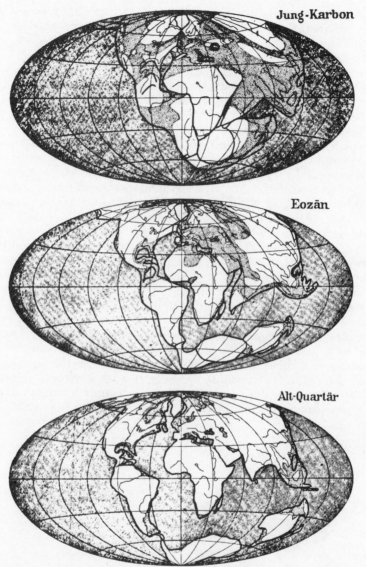

Jung-Karbon

Eozän

Alt-Quartär

Wegener's map of the break-up of Pangaea in the Late Carboniferous, Eocene and Early Quaternary (from the 3rd German edn)

The Francophone community obtained their Wegener in 1922. The 13th International Geological Congress had been allotted to Belgium at the 12th IGC in 1913 at Toronto. The German attack on Belgium caused the event to become delayed for six years, eventually being organised for August 1922. It was to be a symbol of regeneration from the ruins of Europe. Yet the agony of the Kaiser's war lingered; all Germans were excluded from participation or attendance at the Congress and there was no mention of Wegener in the conference programme. The mediation, in a time-honoured manner, came through the Swiss.

Emile Argand was Wegener's contemporary, born in 1879 in Geneva. He had fought off an early apprenticeship with an architect, organised by his father, and a series of attempts by his mother to direct him towards medicine, to allow his love of mountains to be consummated in Geology. It was as an architect that Argand approached the Swiss mountains, visualising the structures of the Alpine nappes with considerable spatial skill and artistry. He could memorise all details of a map and recall the shapes for reproduction. While Marcel Bertrand had recognised the significance of the nappes, it was Argand who sought to explore the total architectural structure of the Alps. In 1915 (following the lead of Einstein) in a lecture to the Swiss Geological Society,[20] Argand announced that the new dimension of time would have to be added to his spatial analysis to permit complete elucidation. The study of evolving mountain structures he termed 'embryo-tectonics'. Argand was also enquiring into the ultimate cause of mountains. This was a philosophical journey rather than an exploration of mechanics; a journey without end – the last part of his life was spent in writing about philosophy and linguistics.

In 1915, in the midst of the war, at a time when all German publications were prohibited from entering Switzerland, Argand gained an illicit copy of Wegener's book and so learnt of the theory of continental displacement. Formerly Argand had seen the Alps through the vision of Suess but suddenly the large-scale horizontal relocation of the nappes could be explained, not as a troubling exception in the gradual contraction of the globe, but as a universal phenomenon.

Emile Argand, 1879–1940
Swiss geologist

Argand gave to Wegener's theory the same passionate and enthusiastic support that he gave to unravelling the Alps. Through the moving continents he projected his work into the other great mountain ranges of Eurasia. The clarity and vigour with which he interpreted the Alps, now capped by this new universality, gained for Argand the prestige of being invited to give the inaugural address in French at the Brussels Congress.

Exactly what he said in this lecture remains mysterious, for the written account[21] in the Congress Proceedings occupies 200 pages of an almost philosophical treatise on the origin of mountains. It begins with his own location within the history of Geology: 'Twelve years have passed, gentlemen, beyond the last pages of the great work of Suess'. Three-quarters of the way through, at Part 26 of the text, Argand, in following the course of his own mental progress, that itself has followed the evolutionary development of the mountains, crosses the watershed of description into causality: 'The great mass of information now assembled demands that a new state in the enquiry should be reached'. The territory of Geology runs out – the unknown and uncertain province of Earth theory stretches into the distance. He wanders into *terra incognita* and sees all around him a bewildering display of explanations. Within this pilgrim's progress, he describes his perception of there being a central battle between theories that imply fixed continents and those that demand large-scale horizontal movements. He christens these primal warring powers 'fixism' and 'mobilism'. This was Argand's greatest innovation, to perceive that the basic schism was between two entirely distinct gestalts: the fixist position of the traditional geologist who studied and collected rocks, and that rare and inspired clairvoyance that saw through the substance of the stones to a new scale of the slow movements of the Earth's crust and the great vista of Time.

Monsieur Wegener, Prince of Light, now makes his entrance. Since 1915, states Argand, and especially since 1918, he, Argand, has tested, weighed, investigated, assayed and analysed Wegener's drift theory in every tectonic situation which he has found, both in his own explorations and in the reports of others. And he has not found it wanting. 'Fixism' continues Argand 'is not in itself a theory – it is the abnegation of theory'. The insidious danger of the fixist position is that it inspires in those who support it an inertia of thought. The primary manifestation of fixism, the contraction theory, is totally at odds with isostasy in its demand for the interchange of continental land bridges with abyssal ocean floor.

The rhetoric rises to new heights: Wegener's displacement theory is powerfully supported wherever it overlaps with Biogeography, Palaeoclimatology, Geology and Geophysics. 'Elle n'a pas refutée. . . . On pense tenir une objection decisive, encore un coup et tout va craquer; mais rien ne craque; on n'a oublié qu'un ou plusieurs tours. C'est la resistance protéene d'un univers plastique'. The philosophy of mobilism – of pluralism, plasticity, evolutionism, relativity, movement, liberation and action – is contrasted with the rigidity, dogmatism, authoritarianism, conservatism and stagnation of fixism; life against death, a vital Earth as opposed to a morbid planet.

There was nothing cautious in the support; Wegener was just the latest in a dynasty of Earth metaphysicians: Elie de Beaumont, Lyell, James Hall, Suess, Marcel Bertrand. Argand ended on a note of sublimity – mobilism was part of that metaphysical view that gave to ideas and the imagination a greater and more powerful reality than that of mere substance. Wegener's theory is at one with great art and the finest constructions of the imagination.

It seems implausible that Argand could have delivered such rhapsody, yet as Wegener had provided Argand with *the answer* he could hardly be excluded from the oral version of the dissertation. Whatever Argand chose to suppress in his lecture, tough-nut, rock-hardened geologists must have found all this difficult to swallow. Here was the opening lecture of the first Geological Congress since the darkness of the Great War had fallen across *Das Antlitz der Erde*, and it had turned into a panegyric. Wegener had been launched, but in a strange craft – praised as some kind of messianic figure who had found the key to the the geological mysteries. If the rhetoric was overblown, then, as Argand had turned mobilism vs. fixism into a political and philosophical battleground, so the response could follow a well established pattern of reaction. Although there were a number of British and American geologists present who were to play an important part in the ensuing debates, how many of them could understand French?

Immediately before the Congress, another French Swiss geologist, Elie Gagnebin, having seen the future programme of the Congress, bemoaned the absence of any comment on Wegener, and offered to the French their first detailed account of Wegener's theories.[22] Gagnebin attempted to sell the theory on account of its fashionability: 'In France, where the theories of Einstein and the psychoanalysis of Freud are so wildly "à la mode", Wegener's hypothesis has been almost completely ignored.' Two years later Gagnebin's hard-sell had evidently proved successful, for Jacques Bourcart could write of the fashion-conscious French that 'Thanks to the translations Wegener has become almost as "à la mode" as the psychological theories of Freud, or Einstein's relativity'.[23] As a novelty fresh from Germany, Wegener's ideas were imbued with the spiritual powers emphasised by Argand that also haunted the French love of the anti-materialism offered by Einstein and Freud. Geologists perceived and flirted with the ideas in this light; in 1925 the director of the French Geological Survey, P. Termier, who had provided the Foreword for the French translation of Suess's *La face de la Terre*, wrote that Wegener's concept is 'a beautiful dream, the dream of a great poet. One tries to embrace it, and finds that he has in his arms but a little vapour or smoke; it is at the same time both alluring and intangible'.[24]

The English welcome

Wegener's theory first appeared in England a decade after its emergence in Germany; the war had provided perfect intellectual insulation. The response to Wegener in England was crucial, because of the destiny of the intellectual life of Britain relative to that of Germany. The argument also reached a scientific resolution not achieved elsewhere. In contrast to Germany, British Earth Physics was less a science of observations than of theory, one that had traditionally attracted the interests of some of the ablest mathematicians. For while the debate, wherever it moved, assessed the confused and formless evidence of similarities between the continents, the mathematician could ignore the detailed fortification and aim direct for the theory's soft underbelly, that part that called upon physical explanation. Wegener could be disbelieved and discredited on account of the evidence, but he could not be refuted in any scientific way except by the argument based on physics. The debate in England smouldered for about six years but was effectively all over in two.

The English were introduced to Wegener in the spring of 1922. In 1919 the young Professor of Mathematics at Manchester University, Sidney Chapman, a brilliant researcher in the fields of solar physics, earth magnetism and the behaviour of the upper atmosphere, had been engaged on a vigorous walking tour of Norway when he had dropped in on the great Norwegian meteorologist, Jakob A. B. Bjerknes. A conference of Norwegian and German meteorologists was being held and Chapman was privileged to hear Wegener talk of his theory of continental drift. The Englishman was captivated, and on his return to England, while walking on the Derbyshire hills, expounded Wegener's ideas to his companion W. Lawrence Bragg, the Cambridge-trained crystallographer, then the Manchester Professor of Physics. Bragg later recalled 'Chapman's description of the theory impressed me so much that I can remember the exact spot where he began to talk about it and where the idea of the very great movement of the continents involved in the theory dawned on me.'[25] Bragg was 'so thrilled' that he wrote to Wegener for an account of his theory, and had it translated. It was presented at the Manchester Literary and Philosophical Society meeting held on the spring equinox of 1922[26] by a young geologist, W. B. Wright, and a botanist, Professor F. E. Weiss. Bragg assumed that his audience would be enthralled. Instead the response left him dumbfounded: 'the local geologists were furious; words cannot describe their utter scorn of anything so ridiculous as this theory'.

Shortly before this meeting an anonymous review of the second edition of *Die Enstehung der Kontinente und Ozeane* had appeared in *Nature*.[27] It begins by identifying the German reaction: 'This book makes an immediate appeal to physicists, but is meeting with strong opposition from a good many geologists'. The author of this review was almost certainly Bragg. He demonstrates his considerable enthusiasm without grappling with the detail of the theory, ending with an unsubstantiated but prescient claim that 'the revolution in thought, if the theory is substantiated, may be expected to resemble the change in astronomical ideas at the time of Copernicus'. The English could begin to learn more in a translated article by Wegener, probably sponsored by Bragg, to be found in a spring issue of the magazine *Discovery*.[28]

Bragg the physicist could not hope to deflect geological opinion, for geologists were still living in the era of liberation that followed on the defeat of Kelvin. In 1906 the geologist T. Mellard Reade had written 'the bugbear of a narrow physical limit of geological time being got rid of, we are now free to move in our own field of science.'[29] The new balance of power even gained recognition from the physicists. In a 1922 lecture titled 'The borderland of Astronomy and Geology' given to the Geological Society of London, the renowned astronomer Arthur Eddington stated 'the time has gone by when the physicist prescribed dictatorially what theories the geologist might consider'.[30] In August 1922 Philip Lake, a Cambridge geographer, but more significantly an active geologist and fellow of the Geological Society of London, wrote in the local Cambridge *Geological Magazine* an eight-page critique of drift, which provided the first reasoned attitude of the geological establishment, accusing Wegener of unimpartiality while admitting that the theory was seductive.[31]

The first English debate on drift came at the September meeting of the British Association for the Advancement of Science, held at Hull. The topic was introduced by a geologist, Dr J. W. Evans, recently returned from the Brussels Geological Congress, who was to prove drift's most senior British advocate. The meeting contained an illuminating variety of responses: Dr G. C. Simpson remarked cryptically that 'the theory was a wonderful one from the meteorological point of view'; Professor Turner is critical of the astronomical observations used to verify Greenland's flight across the lines of longitude, but considers such evidence to be the only worthwhile part of the theory; and Professor Gilligan used one speculative whole-Earth theory to damn another, claiming that 'the time-honoured conception that the earth shows a tendency towards a tetrahedral form was in conflict with this new hypothesis'. For the younger geologist, Mr W. B. Wright, who had introduced the Manchester meeting, there was great enthusiasm for the geological compatibility of a rejoined Atlantic. When writing up the meeting in *Nature*, Wright summed it up as 'lively but inconclusive' and repeated the judicial statement of authority echoed from the Germans 'the surest test of its validity lies in the domain of geology'.[32]

The great interest in his theory had prompted Wegener to bring out a third 144-page edition of his *Die Enstehung der Kontinente un Ozeane*, and this was lengthily reviewed in *Nature* in early December, shortly before Wright's report of the British Association meeting appeared.[33] The tone is one of great flippancy; its author, G. A. J. Cole, writes in the priggish humour of *Punch*. Cole begins 'However much conservative instincts may rebel, geologists cannot refuse a hearing to Alfred Wegener, professor of meteorology'. It was always important for a critic to give the outsider status of Wegener prominence; here even in the first sentence it has been made clear that Wegener is writing in a science for which he is not qualified. The theory is humiliated through distancing the critique; the continents are described in a spirit of whimsy as 'dancing'. With the arrival of the report of the British Association meeting only a few weeks later, Cole must have realised that his insouciance had been ill chosen. For at least by a small group of scientists, drift was being taken seriously.

This became self-evident at the most rewarding English discussion of drift which took place at the Royal Geographical Society in London on the 22 January 1923.[34] The meeting was introduced by Philip Lake with a talk that maintained the aggressive style of his earlier review. The fascination Lake evidently had, and continued to sustain for some years to come, in his quarry was that of a cat that plays with its prey for so long that the prey is never consumed. Lake's comments are, however, far less interesting than those of the audience. There is a mood of sympathy, in particular among some leading visiting geologists. First Mr G. W. Lamplugh: 'Wegener has struck an idea that has been floating in geologists minds for a long time'. Then R. D. Oldham: 'the important question is not whether Wegener is right or wrong, but whether the continental masses have throughout all times maintained their present position.' Mr F. Debenham: 'In justice to Wegener I think we ought to realise that he is an exceedingly bad advocate for his own theory. . . . We imagine we know all about the ordinary processes of the minor features of the Earth . . . but larger areas have been somewhat neglected'. Mr C. S. Wright, directing his remarks even at the man who introduced the meeting: 'With a hypothesis that touches so many sciences the only effective attack is a criticism of the hypothesis as a whole, not on a few points of detail'. The wise intellectual calm of this meeting was never to be repeated.

The meeting crystallised a widespread attitude of benevolence. Continental drift became fashionable in the winter of 1922–3. The new success of a British popular press had invented the image of the mad foreign scientist as exemplified by another German Einstein. As in France this was an era prone to vogues; air-travel around the Atlantic had given rise to 'Atlantic-fever'. Both Meteorology and continental drift could feed off this interest. There was a feeling abroad among young geologists, a feeling that would not reappear until the 1960s, that the old order had had its day and that such a new theory could hope to flush out the 19th-century backwaters of Geology. Students caught a new wider cosmological scope to their own subject and many of them willed it to be true. The popular success of continental drift can be gauged by the attention given to it within the *Nature* letter columns; for within these first few months it is seen as offering the potential to unify puzzling observations made from across the natural sciences: from the life-cycle of the eel in its journey to the Sargasso Sea, to the rotation of the pyramids of Cheops by 4 minutes of arc since the time of their construction. Such insights passed unnoticed by Wegener for on 23 May 1923 *Nature* announced that the calamitous depreciation of the currencies of Germany and Austria had prevented those countries obtaining foreign scientific periodicals.)

The new climate of sympathy gained support from the critics themselves who had been making errors, thus revealing their true motives, less to assist in the production of good science, more to see this theory destroyed. In his introductory talk to the Royal Geographical Society, Lake quoted the work of one G. V. Douglas to show how the two distinct elevations of the surface depth-heights of the Earth could result from the deformation of a single type of crust into broad sinusoidal folds. Douglas's paper subsequently appeared in the *Geological Magazine*[35] where it was peremptorily exposed by the young self-appointed defender of drift, W. B. Wright,[36], as requiring every valley and every plateau on Earth to be of the same respective elevations.

The most important reaction to continental drift, was however that of institutional geology. The two principal London societies, the Geological Association and the Geological Society of London, the latter being the first and greatest geological society in the world, neither discussed nor published any article on the subject of continental drift during this first phase of enthusiasm. The whole-Earth theory was not even worthy of dispute.

The man who was later, after the 1960s claimed to have been the most influential of the whole debate had been doggedly attending the discussions. Harold Jeffreys was a young Cambridge mathematician; the natural heir to Kelvin. It was Jeffreys who would reintroduce the

Sir Harold Jeffreys, 1891–
British mathematician and Earth physicist

reign of the physicist, after its 20-year interregnum, only this time geological orthodoxy would find itself in alliance with its former antagonist. Jeffreys' response became gradually sharpened. At the British Association for the Advancement of Science meeting he gave only disconnected criticisms: that the rotational force is insufficient to explain the crumpling of the West Coast American Pacific Ranges; and that ocean floors, being less radioactive than the continents, should also be stronger. At the Royal Geographical Society meeting, he is becoming more forceful: 'the physical causes that Wegener offers are ridiculously inadequate'; but then, at the beginning of 1923, still recognising the power structure after the Kelvin débâcle, he admits deferentially 'I think that there are probably two forces which are perfectly capable of producing large migrations of land and sea, if they are wanted': a change in the Earth's speed of rotation, or if early in the Earth's history all the land had ended up in one hemisphere, then it would tend to redistribute itself more evenly across the globe. Jeffreys is actually offering a lifebelt to continental drift, but only if it abjures those forces advocated by Wegener.

By April 1923 in a letter to *Nature*[37] this lifebelt had been withdrawn. He is reasserting the transcendence of an argument based on mathematical physics: that the very existence of mountains proves that the crust of the Earth does not readily yield to small forces, even when operating over long periods of time. Under such circumstances mountains would flow away.

In 1924 Jeffreys wrote a book, boldly titled *The Earth* that was the product of a course of lectures, 'The Physics of the Earth's interior', he had been delivering since 1920 at St John's College, Cambridge. The physicist's authority is stated within the introduction: an 'objection I anticipate is that the book is too mathematical for geological readers, but if geophysics requires mathematics for its treatment, it is the earth that is responsible not the geophysicist.'

Towards the end of 1924, 18 months after the original enthusiasm, the 3rd edition of Wegener's book was finally published in an English translation. A tentative and weakly argued review of the work, written by J. W. Gregory, appeared in *Nature* in February 1925, claiming that Wegener had gained 'a favourable reception'.[38] The reception, however, was almost over. In April 1925 an American geologist, A. P. Coleman, wrote in the columns of *Nature*[39] 'that a careful study of the two greatest periods of glaciation known to geology gives no support to the theory of the drift of the continents'. A month later in *Nature*, E. Meyrick[40] analysed the supporting evidence for the distribution of Microlepidoptera and stated 'that if Wegener's views on Australia and New Zealand are an integral part of his scheme, the hypothesis is disproved by facts'. Such attacks against the central climatological and biogeographic arguments on which Wegener gave such credence were extremely damaging.

Philip Lake had a final opportunity for lambasting Wegener in the *Geological Magazine* where he reviewed the English translation with a smug xenophobia claiming that 'the printing and paper are far superior to those of the German publication, and while the latter is almost repellent in form, the English edition is quite an attractive volume'.[41] Lake's tone was at least representative. No one came forward to speak in Wegener's defence; the scientific response had tilted to the generally unfavourable; those who had been converted to a broader mobilism preferred to keep their peace.

One reason why the publication of an English *The origin of the continents and oceans* did not inspire fresh enthusiasm was because this, the translated 3rd edition, seemed in several places to be outmoded by the debate. In 1925 J. W. Evans became President of the Geological Society and offered in two anniversary addresses the first sympathetic response to Wegener articulated in this forum, though preferring the version offered by Pickering, that the loss of the Moon out of the Pacific provided the best explanation for any such movement.[42] In 1929 J. W. Gregory became president: 'If isostasy be so stated that it is inconsistent with the subsidence of the ocean-floors, so much the worse for that kind of isostasy'.[43]

The Geological Society finally got round to a drift debate in 1935, 12 years too late, at the unflinching instigation of Mr W. B. Wright and in collaboration with the Royal Astronomical Society.[44] By bringing in the astronomers, Wright had anticipated some broader scientific support. He was to be disappointed. The leading astronomer Sir Frank Dyson stated 'that he did not think that astronomers could say very much about Wegener's interesting hypothesis.' The debate included nothing on the developments since the early 1920s. There was the established support from those whose views were built on Zoo-geography: lemmings attempting a periodic journey across the non-existent North Atlantic land mass and plovers that migrated each autumn from British Columbia to Hawaii now joined the eels in showing their support for drifted continents. The ubiquitous Jeffreys was there to dampen enthusiasms leaving Wright to bemoan that 'the evidence in favour of the theory had not been adequately dealt with either by himself or any of the other speakers'. The astronomers thanked their hosts; grieved that 'modern specialization led too far in causing the different scientific groups to drift apart' and returned across the courtyard of Burlington House to their own Society's home.

The American rejection

Continental drift never had its winter of 1922–3 in North America. Post-war isolationism militated against the adoption of foreign novelties. In 1920 Taylor, inspired by unknown motivations, but possibly just because the Geological Society of America circus had that year come to town, presented to the Chicago meeting two short papers repeating his theories.[45] In October 1923 Pickering in Jamaica picked up rumours of the English revival of continental drift and sent to the

Cambridge *Geological Magazine* an update on his 'The separation of the continents by fission' which was published in 1924.[46] Pickering attempts to demonstrate his own priority in some of the ideas, restating them with a dash of poetry: as the Moon left the Earth 'Three quarters of the Earth's surface, to a depth of 35 miles, was carried away in a trailing mass of ruin. New Zealand itself was just saved to the Earth'.

None of the popular science magazines gave Wegener any attention until after the 1925 English translation. It is telling that Pickering chose to republish his theories in England rather than his native America. The most damning of all the reviews that Wegener was to receive in either England or America also contributed to the apathy. The mathematical geologist, Harry Fielding Reid 'the first geophysicist in America', had in the second volume of the 1906 San Francisco earthquake report produced the celebrated theory of 'elastic rebound' that explained how an earthquake was the product of a sudden release of stress along a fault. In 1922 Reid was invited to write on the German 3rd edition of Wegener's work by the American *Geographical Review*,[47] a journal that, unlike the geological magazines, considered the subject to fall within its ambit. If the English attitude to drift was playful, then Reid, the mathematical geologist, is hardline, displaying not a hint of humour.

> There have been many attempts to deduce the characteristics of the earth from a hypothesis; but they have all failed. . . . Elie de Beaumont . . . Green. . . . This is another of the same type. Science has developed by the painstaking comparison of observations and, through close induction, not by first guessing at the cause and then deducing the phenomena.

Reid's argument had great resonance, for this is geological puritanism, as articulated more than 100 years before at the founding of the Geological Society of London.

The first opportunity American geologists had to test their opinions came at the Geological Society of America's meeting at Ann Arbor, Michigan, in the dying days of 1922.[48] The presidential address from Charles Schuchert on the 'Sites and nature of North American geosynclines', with subsequent papers on parallel themes could not fail to include something of the news from the east. Schuchert is unhesitant in his support for 'earth shrinkage', so it was left to a subsequent speaker, A. Keith, to include both Taylor's and Wegener's theories in his review of mountain-building explanations. Taylor merits four pages but is dismissed as requiring 'extravagant motions of enormous masses' with a mechanism which 'meets an insurmountable obstacle in friction'. Wegener, author of 'the latest form of continental creep', is more easily dismissed in only two pages.

The only speaker prepared to mount a defence for even the broader scope of Wegener's theories was Reginald Daly, who, unique among

Reginald Daly, 1871–1957
American geologist and petrologist

American geologists (Taylor excepted), was a mobilist. Daly was a widely travelled and imaginative synthesist in the mould of Fisher and Dutton. Trained as a mathematician, he had become interested in Geology at the age of 20 while working as an instructor at the University of Toronto. He was now Professor of Geology at Harvard. In contrast to the normal geological concern for field mapping Daly's mathematical training had taught him to search for problems awaiting solution. He was particularly interested in igneous rocks and active volcanoes; his line of enquiry went direct to the root of the problem – the source of the magma.

The first few years of his Harvard professorship were crammed with field expeditions; to the volcanic islands of Hawaii, Samoa, St Helena and Ascension. In 1914 he wrote a book titled *Igneous rocks and their origins*. Like Green he was constructing for himself a confined world, everywhere possessed with the rapid potential to bring forth magma. Like Fisher, he saw in the increase of temperature with depth that below a solid crust of only 40 km thickness the rocks should be molten. His answer to the absence of earth tides and the passage of seismic shear waves was to presume that under the high confining pressures this magma had the physical properties of a glass. The glass was a universal substratum. Such a material could provide the lubricant to permit the horizontal movement of continents.

Daly's opinions were received with great scorn at Ann Arbor. Professor W. H. Hobbs employed the familiar denunciation of the theorist as an advocate: 'I have thought if he were not such a brilliant geologist what a fine trial lawyer he would be. I have been impressed not with the inadequate attention the Wegener theory has received, but rather with the exaggerated welcome that has been accorded it'. Professor A. P. Coleman from Toronto who had originally inspired Daly to take up Geology, next joins the censure, blessing his protégé's 'poetic imagination'. (It was Coleman who three years later was to publish in *Nature* an article opposing drift with all the venom of the North American condemnation.) This was the closest the Geological Society of America came to discussing drift.

The first formal meeting on drift came four months later in April 1923 when the Philosophical Society of Washington invited the Geological Society of Washington to a joint meeting consisting of three addresses on 'The Taylor–Wegener hypothesis'.[49] The three speakers were Taylor, Daly and W. D. Lambert of the US Coast and Geodetic Survey. Taylor provides his usual story, the eternal circle of reason in which the capture of the Moon at the end of the Cretaceous is used to justify the sliding continents, and the sliding continents used to justify the capture of the Moon. Daly shows more reserve than at Ann Arbor and now offers a 'Critical review of the Taylor–Wegener hypothesis' that contains a cautious welcome to horizontal movements wrapped within a scepticism more compatible with his compatriots. Lambert feels qualified to speak only on the forces producing the movement but 'not being a geologist' provides a tactful commentary. He declines to comment on Taylor's demands for an arriviste Moon, but concludes in respect to Wegener 'till some more adequate explanation is offered mathematicians and physicists are likely to doubt the validity of the hypothesis.' End of meeting.

These two informal discussions gained no outside publicity and provide the totality of the early American response. The choice of speakers in Washington shows a deliberate attempt to obtain a sympathetic hearing. There was no substitute available for either Daly or Taylor to speak in favour of mobilism. The first intimations of continental drift that the majority of American scientists were to receive came in the reviews of the English translation of Wegener's book. In *Science* for July 1925[50] the reviewer notes the considerable attention the theory has received in Europe, presents certain details without enthusiasm, and wheels out the by now established critique. *Scientific American* of January 1926[51] attempted to whip up some journalistic froth over Wegener: 'The theory is startling. To many it seems absurd. It may prove to be erroneous. It may gain final acceptance among geologists.' The article continues in this mix of Hemingway and equivocation: '. . . the lunar method [of geodesy] is not absolutely precise. Can these figures be trusted? In any case they will not be trusted.'

The English backlash to Wegener had already taken place; Jeffreys' *The Earth* had been published in advance of the theory he wished to condemn. While in England the backlash never entirely destroyed an undercurrent of support, in America the new wave of criticism had arrived to fortify the pre-existing hostility. The difference between the British and the American response was fundamental; the divergence in the welcome extended during the winter of 1922–3 became crystallised into attitudes towards mobilism that maintained themselves through into the 1960s. The Americans were almost universally hostile; in Britain there was always a significant proportion of scientists in broad sympathy.

The last valiant attempt to mount an evangelical mission on behalf of drift was organised by a visiting Dutch geologist in 1926, but succeeded only in writing in a widely circulated book, drift's obituary.[52] This notorious meeting provided the greatest cast of any of the debates, yet as with so many pieces of elaborately staged drama, there was an overt desire to role-play and exchange was frozen. The Royal Geographical Society meeting three years earlier had provided a far more intelligent discussion, but at that time the whole subject was fresh. It was now exceedingly stale. In place of insight the meeting highlighted prejudice; Wegener was a heretic to be drummed out of town, and had mistakenly crossed the Atlantic to be so abused. This was his first and only visit to America and encounter with Taylor with whom his name was in that country so often married.

The organiser of the meeting was W. A. J. M. van Waterschoot van der Gracht, an eminent and enlightened Dutch oil geologist who had been living in America since 1917. At the time of the meeting he was the Vice President of the Marland Oil Company, and it was with such influence that he obtained a meeting under the patronage of the American Association of Petroleum Geologists.

Besides the protagonists, Wegener and Taylor, the meeting contained a curious mix of geologists, who offer two distinct styles of attack. The first of these was exemplified by Reid's 1922 review: that the enterprise of constructing imaginative whole-Earth theories is in essence pseudo-scientific. This was the form of the 19th-century argument. In the new 20th-century form, as articulated at the meeting by Rollin T. Chamberlin (the son of the aged T. C. Chamberlin, who his father, like a great patriarch, had seen fit to install in his own Chicago department), there already was a perfectly satisfactory whole-Earth theory in existence: his father's planetissimals. There was thus no need for this intruder, this foreigner, especially when Chamberlin senior had already made the acceptance of his theory a prerequisite of patriotism. While geologists in England, who had produced no whole-Earth theories to improve on the contracting Earth, found it easier to borrow from Wegener, in America most geologists adopted a position somewhere between the arguments of Reid and Chamberlin. If pressed they might claim, as geologists had always claimed, that such concerns were beyond the needs of Geology.

Within the conference write-up van Waterschoot van der Gracht had the luxury of being able to enclose the proceedings within his own comments; his opening speech was extended to 75 pages; and he devoted 30 concluding pages to a commentary on those remarks made by other speakers. His theme is to make Wegener's hypothesis work, through incorporating the idea of other mobilists, and to play down the greatest internal contradictions: 'It will take generations before everything can be explained and every question answered'. In the addresses that followed, while Taylor repeated his theory in full,

Wegener limited himself to two details, first concerned with attacking the evidence for Coleman's proposed 'Permian American ice sheet', when that continent had, according to Wegener, been lodged in the tropics; and secondly providing additional information on his firmly held belief that measurements of changing longitude would provide the only proof of drift. These two details seem to be no more than footnotes. Significantly, Wegener's lack of English left him unable to refute his accusers. Van Waterschoot van der Gracht, who had spent several years in Germany, acted as his interpreter for the address but could not hope to encourage spontaneous discussion.

The critics were in general gentlemanly in their distrust. Two speakers delighted in playing games according to rules set down by Wegener: Professor Chester Longwell from Yale illustrated how the continent of Australia fitted snugly into the Arabian Sea just as Africa locked with South America; and Charles Schuchert, emeritus Professor of Palaeontology also from Yale, displayed the ludicrously mismatched results of shifting Plasticine continents across an 8 inch (200 mm) globe. (Schuchert had rotated them around the North Pole.)

The most savage response came from Chamberlin's son, who indulges in sarcasm and vitriole at the expense of the mute Wegener. He begins 'Wegener's theory, which is easily grasped by the layman because of its simple conceptions, has spread in a surprising fashion among certain groups of the geological profession. Other groups of the profession ask: "Can we call geology a science when there exists such difference of opinion on fundamental matters as to make it possible for such a theory as this to run wild?"' He proceeds to list objections, claiming that 'Wegener's own dogmatism makes categorical comments somewhat less objectionable than would otherwise be the case'. The introductory invective is followed by the list. The style is violent. By the end of objection 15 Chamberlin's imagination has dried up but his anger sustains him into objection 16: 'A great deal of Wegener's argumentation seems very superficial, and the facts involved seem to the writer in many cases to point to very different conclusions from those utilized by Wegener' and objection 17: 'Wegener's hypothesis in general is of the foot-loose type, in that it takes considerable liberty with our globe, and is less bound by restrictions or tied down by awkward, ugly facts than most of its rival theories. . . . The best characterization of the hypothesis which I have heard was a remark made at the 1922 meeting of the Geological Society of America at Ann Arbor. It was this: "If we are to believe Wegener's hypothesis we must forget everything which has been learned in the last 70 years and start all over again."' The final objection, objection 18, is closer to home: 'The planetissimal hypothesis, instead of being detached and free-floating, is an integral part of a comprehensive geological philosophy. Wegener seems to be entirely oblivious of its existence.'

Chamberlin's comments were to resonate long after the debate was over as geologists in America read the 1928 write-up of the symposium proceedings to assess the balance of argument. They would find that van Waterschoot van der Gracht's attempt to rewrite Wegener in a more generous mobilism had the effect of undermining the original author; for to attempt to save his theory through the construction of apologia seemed to provide a hollow, ambiguous defence.

The most extraordinary feature of the book was not the text but the index. The strangest items have been included: 'Co-operation in approaching truth, a feature of symposium on continental drift' refers the reader to page 197. Most tellingly, under the letter 'O' there is 'Objections to Wegener's theory of continental drift, followed by a list of the individual objectors with page numbers: 'by Berry, 194; by Bowie, 178', etc. There is no entry under 'Favored by' and instead as a subheading beneath 'Wegener's theory . . .' those who support the theory are listed above a repetition of the list of objectors. This simplistic reduction of a complex debate has all the makings of a vote. Discounting Wegener and Taylor as the defendants and van Waterschoot van der Gracht as the self-appointed judge, there are 12 strong men left as the jury. The courtroom drama has been brewing over the years. The index spells it out, listing 'Wegener's attitude, that of an advocate of his theory rather than that of a scientific investigator' with the names of three 'jurors' who testify to this opinion. Guilty or not guilty? Of the four in favour of the theory, three of their names also appear under the list of objectors. Therefore the final poll would appear, according to the logic of the anonymous indexer, to be eight to one with three undecided. Not enough to hang but sufficient to banish to the antipodes.

The death of Wegener

The 'drift' that the English-speaking world debated was that of Wegener's 3rd edition. At the 1926 New York meeting Wegener was offered the opportunity to reveal exactly how far his opinions had subsequently evolved. The opportunity was declined. He was beginning to feel overwhelmed by the scale of the controversy, and the quantity of foreign language comment, as well as astonished by the American hostility.

Wegener loathed the bureaucratic responsibilities entailed by his post-war position as head of a research organisation. In the absence of any offers from German universities to provide a chair for this eclectic scientist, the University of Graz in Austria, where his brother already taught, eventually created a personal professorship. In 1924, with his wife, their three children and the 78-year-old grandmaster of Meteorology, father-in-law Köppen, Wegener moved into the Austrian hills. Freed from administrative chores he was an enthusiastic teacher. His biographer Benndorf writes of him 'it was the simple and unpretentious manner . . . that took the hearts of the young people by storm. . . . I think that they would have gone through fire for Wegener; and if anyone had dared to doubt the theory of continental drift they would not have hesitated to use their fists as an argument.'[53]

This adulation did not help Wegener restructure his arguments; in November 1928 he had finished the proofs of the 4th edition of his *Die Enstehung der Kontinente und Ozeane*. Despite the wealth of constructive criticism that his proposals had received over 16 years, through four editions of his book, Wegener persisted with certain incommensurabilities and anachronisms. Never during this period had he attempted to deconstruct his original proposals and observe their component parts in order to rebuild his theory. Instead, each article, each book, took an identical framework and simply sought to incorporate further substantiation. The most famous example of this was Wegener's insistence that the glacial moraines of Europe and North America had been conjoined until the North Atlantic had begun to open a mere 100 000 years ago, a date that compared with the 100 million year age of the Sierra Nevada and Rocky Mountains, formed, he claimed, as North America trundled into the crust of the Pacific. The corroboration that he sought for such rapid opening lay within the astronomical observations of longitude, measurements that he well knew could not be assured for accuracy. With such rates of movement, oceans could open and close in a blink of a geologist's eye; and yet the disintegration of Gondwanaland had taken more than a 100 million years. In reading Lyell, he could have found the rates of geological processes as revealed by the fastest rising coastlines of the earthquake-prone Andes: a few metres per century. A closer scrutiny of Wegener's arguments reveals that they are reasoned, ordered, until they begin to approach his beloved Greenland. Throughout his life Greenland had been at the centre of his thoughts as that subcontinent had provided the keystone of the North Atlantic. The emotional magnetism of this strange land perpetually warped his rational insights.

Yet the arguments of the 4th edition are everywhere presented more soberly and scientifically than in any earlier volume. Enthusiasm is subdued: 'The Newton of drift has not yet appeared' he claims in a mitigation of the absence of a motor to drive the continents. 'His absence need cause no anxiety'. This was a singular deception. Despite a brief and uninspired paragraph on the possibilities offered by internal convection currents, Wegener was deeply worried. He informed his brother that no more updates of the book could be contemplated, and had already begun the planning of a new great expedition to Greenland.

In 1926 and 27, his former student and friend, Johannes Georgi, while working in the far north-west corner of Iceland, had discovered an intense, high altitude stream of air, the jet-stream, passing from Greenland and heading direct for Europe. Georgi, wishing to learn more of the origin of the jet-stream, wrote to his former tutor 'as the greatest expert on Greenland' to seek support for an expedition to man a winter research station at the centre of the ice-cap. Wegener responded enthusiastically, recalling a similar ambition of his own.

His desire to escape from Europe; his comment 'I too have had my own trouble and am no longer a young man'[54] suggest a depression had come over the meteorologist which he hoped a return to Greenland might dissolve.

Such a depression could explain that, although his preoccupation with the westerly drift of Greenland became familiar to his colleagues on that expedition, as leader he made no attempt to plan for research into this movement, despite having with him an experienced geodesist. The 'Newton of drift' had yet to descend out of the clouds. The expedition, originally planned to be in collaboration with Lange Koch, was postponed on the death of his Danish coleader, and became a solely German affair. After a preliminary visit in 1929 the main expedition left in April 1930, but was delayed by the late melting of the ice pack. Huge quantities of supplies had to be pulled by dog-sledge to the 'Mid-Ice' scientific station. New ice-thickness determination equipment, that measured the time for echoes to be returned from surface explosions, had also to be tested. Before the scientific station could be properly constructed, winter was upon them. On 21 September Wegener, with 13 Greenlanders and 15 sledges, set out from the coast with his German companion Fritz Loewe to bring 1800 kg of supplies to the two beleagured meteorologists at the ice-cap's centre. After 160 km of terrible blizzards all but one of the Greenlanders turned back, and so Loewe and Wegener continued with the most essential stores: Christmas parcels, gasoline, scientific instruments and a gramophone. They arrived on 30 October having had to abandon even these supplies; Loewe was suffering from frostbite and would have to stay the winter. The food intended for two would have to be extended for three, but could not feed five. On the morning of his 50th birthday, 1 November, Wegener and the Greenlander Rasmus Willumsen, set off for the coast, with 17 dogs and 2 sledges. It was a 400 km journey; the perpetual Arctic night was closing in around them; the conditions were worsening with temperatures below −50°C, and more blizzards expected.

Exploration in Greenland still held fascination, particularly in America where the *New York Times* provided regular coverage of the expedition. At the coastal station when Wegener did not return there was the assumption that he had chosen to winter at 'Mid-Ice'. With no radio to contact the outside world the ice-station scientists believed their leader to have returned safely. The following spring Wegener's body was discovered, buried in a shallow snow grave marked by his skis, about half-way to his destination. He had died of a heart attack, presumably in his tent, at night, as the result of exhaustion. His Eskimo guide had removed a few items, the last crucial diary, his pipe, and had travelled another 15 or 25 km before vanishing into the oblivion of the drifing snow; his body never to be found. A suggestion went round in the University of Graz that Wegener had been murdered, but this was rapidly discounted. The speculation that had been voiced about his safety all winter now climaxed. Wegener died as a hero. The German government bestowed on the expedition a great national status; offering to take his body home on a battleship and accord him a full military funeral. His family preferred otherwise; Kurt travelled to Greenland to take over the running of the expedition, and on the instructions of his wife Else the body was left where it was found. Perfectly preserved in the deep cold of the inland ice, each year becoming further entombed by the winter's snowfall, Wegener's body will remain in the continent that he loved, for thousands of years until the cryogenic spell is broken and he is cast in an iceberg to drift across the North Atlantic.

5

The war of the worlds

The hot Earth

Wegener's drift theory was an important and early symptom of a malaise spreading through the geological sciences. In a time of transition, as unity begins to dissolve, so it becomes impossible to take a single thread; to perceive the character of a period in a linear history. The weave is tangled; after 1910 came the 'Middle Ages' of Earth understanding. The beginnings of the disintegration of an old order had yet to throw up an effective replacement. At such times archaic theories become resurrected; insubstantial novelties are elevated to serve as central beliefs; as debates become more trivial or more elusive, so the rhetoric is more strident. To gain a grasp on this period it is politic to begin with its most influential success: the study of Earth heat.

The American geologist Daly had been the sole respected figure in that country to support a moving crust. His studies on volcanoes and the Earth's heat flow had convinced him that the material beneath the crust must be in a glassy state, too hot to crystallise. The other two influential English-speaking 1920s mobilists, Holmes and Joly, like Fisher and Green before them, were also inspired by heat, by the Earth's volcanoes and geothermal gradient. Like Fisher, all three had arrived in Geology from pure Science; from Physics, Chemistry, Mathematics. Each proposed Earth models that were contrary, and incompatible; they represent a 'school' only in their opposition to the orthodoxy. The Earth had always been what the imagination made of it. To the investigators of heat it was peculiarly warm.

John Joly, born in Ireland in 1857, graduated from Trinity College, Dublin, with a degree in Engineering, Physics, Chemistry, Geology and Mineralogy and subsequently became assistant to the Professor of Engineering, and then from 1897 the Professor of Geology. From 1899 he was involved in the age of the Earth controversies when he obtained the figure of 80–90 million years since the condensation of the atmosphere, from an estimate of the time necessary for the world's

John Joly, 1857–1933
Irish geologist

rivers to have provided all the dissolved salt in the oceans.[1] This figure was peculiarly credible; it was derived from geological information and therefore demonstrated a new-found quantifiability of that science, but also, most importantly, did not contradict Kelvin's own estimate.

The discovery of radioactivity radically altered the direction of his research. In a letter to *Nature* in October 1903,[2] Joly was the first geologist to proclaim the significance of an Earth rich in radioactive atoms. Yet while Joly was anxious to make Geology compatible with Physics he was also convinced as to the truth of his earlier estimate of the Earth's age. By 1906 it was becoming clear that lead was the end product of a chain of radioactive decay beginning with uranium, and that helium was a product of a variety of radioactive elements, in particular the first-found radium. These results had opened up the possibility that rocks could be dated by measuring the relative proportions of some parent radioactive atom and its daughter decay product. The first ages to be so obtained ranged up to hundreds of millions of years.[3]

Joly, along with most other geologists, was unimpressed. Radioactivity had passed beyond the confirmation of Geology's Earth, and was threatening to undermine all previous sober estimates of the Earth's antiquity. In 1904, two German chemists had announced that even garden soil contained radium.[4] Soon measurements of the concentration of radioactive elements in surface rocks had shown not that the temperature gradient was too great, but that it was far too small if this concentration was in any way typical of the whole Earth. Physicists had been wrong before and would no doubt be wrong again. The geologist Sollas wrote in 1905 that radium was an apparition 'vague and gigantic, threatening to destroy all faith in hitherto ascertained results, and to shatter the fabric of reasoning raised upon them.'[5] Of all the opponents of radioactive dating the most influential was John Joly.

In 1909 he wrote a book that addressed this theme titled *Radio-activity and Geology*. Geologists had long asserted their independence and now, just when the battle with the physicists had finally turned, Joly wished once again to deny that Physics could overrule his science. His own age estimate from the salt concentration of the oceans was *geological*; dating by radioactivity required the extrapolation of a dangerously novel set of principles. Yet while he doubted that radioactivity had allowed a senile Earth to exist in a state of suspended juvenility, Joly was still fascinated by the new phenomenon, promoting the purgative powers of radiation in the curing of cancer. By the 1920s his rejection of radioactive dating was in hopeless isolation, yet Joly still refused to believe that his own age estimates were in error. Like Kelvin, he died without publicly recanting his beliefs.

In order to make all the new evidence for radioactive heating compatible with his own 100 million years, Joly was led to construct a new terrestrial history. The heat surplus provided by radioactivity began to dominate his thoughts, and as his thoughts led him to believe, to have dominated the history of the Earth. Extrapolating from the near-surface temperature gradient, Joly, like the 19th-century fluidists, found that rocks at only a few tens of kilometres of depth should be molten.

Having first delivered a lecture at the Geological Society of London in May 1923, in 1925 he formed his ideas into a book titled *The surface history of the Earth*. 'American geologists designate the mountain-building periods "revolutions"' wrote Joly. His theme is almost anthropomorphic: to explain 'the birth, advent, consummation and decay of world revolutions'. Their origin lies in the occasional and periodic liquefaction of the upper layers of the substratum as the heat generated by the radioactive atoms builds up to this inevitable result. Disaster threatens; the outer crust of the Earth begins to shift

> but just when the substratum melts and further supplies must involve rising temperatures and ultimately melting continents, astronomic interference comes in. The ocean floor is caused to thin out under the fiery tides raised in the underworld; the ocean rapidly absorbs the heat and all danger is averted. The shifting of the earth's crust over the substratum brings the same redemption to the whole Earth.

This was no more than some old-fashioned catastrophism of his youth, rewritten in terms of the fear of meltdown. Hell-fire is about to be unloosed; the celestial Moon brings about salvation. Revolution and redemption; this is the Catholic cosmology given to the young James Joyce.

While Joly held continental drift in some distaste, he considered that within his world revolutions, continents slid about forcing up great mountain ranges where they drunkenly collided. This was therefore an episodic mobilism; an essentially fixist Earth indulged in short-lived bouts of frenzied activity. Mobilists, however, saw in Joly's calculations a way in which the friction at the base of the continents could be lowered through melting, to permit more continuous movements. In his 4th edition Wegener had used Joly's theory in this form, hoping thereby to find further corroboration for drift.

The last of the three Earth heat mobilists was also the greatest intellect. Arthur Holmes, born in 1890, was a schoolboy in Newcastle, England, when radioactivity was first announced. His first interest in Geology came from these schooldays but he went to Imperial College, London, to study Physics under the future Lord Rayleigh, then R. J. Strutt. Although Holmes changed courses to graduate in Geology, it was under Strutt that he pursued research into the significance of radioactivity for Geology. Strutt, almost alone, had been investigating the possibility of using the helium content of a variety of uranium-rich minerals to estimate their age. Having found that the gas readily escaped from the mineral structures he encouraged Holmes to develop the alternative uranium–lead dating method.

While a researcher, Holmes first travelled to east Africa, a region that was to become centre stage for the discussion of rifts and rift valleys. At the southern end of the rift, in Mozambique, he saw in the folds of the hills, skin pulled taut across the underlying bones of the geology – a naked landscape in which the relationship between the geology and the surface features was explicit. From the volcanic rocks he collected Holmes wished to enquire into the processes deep underground that such materials sampled. These interests, generated out of the field

Arthur Holmes, 1890–1965
British geologist

geology of Africa, became for Holmes the foundation of a total study of the Earth. Lucid prose, modesty in the face of grandeur, the use of photography and a truly global world-view ensured that Holmes gained the widest audience of perhaps any geologist in the 20th century.

As a research student Holmes set about mastering and refining the new techniques for measuring the age of a rock by accurately determining the relative concentrations of the parent uranium or thorium and daughter lead atoms. With the knowledge of the rate of decay of the parent to the daughter it was possible to gauge how long the system in the rock had remained undisturbed. Because the pathway from uranium to lead was peculiarly complicated, others had abandoned their researches, leaving the 21-year-old research student to become the world authority on a technique that was finally to provide the planet with its authentic, scientifically determined birthday.

Rare among geologists in possessing a complete command of a scientific vocabulary, Holmes presented himself as geology's first formal critic when in 1913 he produced a review of all the methods then available for measuring geological time.[6] The refutations of Kelvin's Earth were by now familiar; those of the geologist's own arguments were, however, novel. He saw in their assumptions a common error that lay in the understanding of their founders' philosophy. Joly had adopted a geological strategy whereby the uniformity of geological processes was given primacy over the scientific method. Holmes was the first to identify the battle between the scientific method of empiricism, employed within his own radioactive age determinations based on the extrapolation of experimental results, and the geologist's belief in the uniformity of geological processes that was based on no more than an assumption.

This was a significant assault; for under Kelvin's attack geologists had seen the need to make their own science quantitative from within. Now they were being lambasted for basing such figures on feeble foundations. The mathematically trained Cambridge petrologist Alfred Harker, joined in the affray, writing in 1915 'in reading some very fascinating speculations by geologists of high-standing, I have often wished that some obliging mathematician would put forth a small manual of applied arithmetic for the guidance of the workers in the descriptive sciences.'[7]

These were new skirmishes in a perennial conflict between geologists and mathematical physicists. Only those such as Fisher, Dutton and Holmes could hope to bridge these sciences and thereby transcend the futile and often acrimonious warfare. The geologists' redress was familiar – as expressed by T. H. Holland (1868–1947) at the 1914 British Association meeting: 'The mathematician apparently finds it just as easy to prove that the earth is solid throughout as to show by extrapolation from known physical values that it must be largely gaseous.'[8] The first geologist fully to recognise the significance of radioactive dating was Joseph Barrell, Professor of Geology at Yale.

Barrell had been investigating isostasy. In 1914 he wrote a series of papers detailing insights offered into the structure of the Earth from observations of the recovery of the elevation of the northern land masses after the melting of the ice-caps.[9] The outermost shell was, he determined, very strong and tens of kilometres thick. This he termed the lithosphere. Beneath, there had to be a 'thick, hot, rigid, but weak shell, the asthenosphere, or sphere of weakness'. While pondering the ramifications of isostasy, Barrell became convinced that there were rhythms or cycles of geological change that were so significant as to make the simple averages used by uniformitarians in their estimates of geological time quantitatively meaningless. The present, for example, was a period of extraordinarily rapid erosion, perhaps at rates 10 or 15 times greater than the average. Hence any calculations based on geological processes, such as those to do with sediment accumulation or salt concentration, would prove to be considerable underestimates. This argument provided the all-important escape clause that allowed geologists to reconcile their own estimates of time with those of radioactive age dating and thereby establish 'the magnitude of the framework into which the geological picture must be set.'

Barrell had written the new concordat between the physicists and geologists. Two years after his premature death radioactive dating received official recognition: the 1921 British Association meeting saw a discussion at which the technique was endorsed by figures familiar to the drift debates, including Harold Jeffreys and J. W. Gregory. In 1922 a meeting organised by the American Philosophical Society had arrived at similar conclusions – the greatest defence of an elder Earth coming from an elder geologist, Thomas Chrowder Chamberlin. Holmes, now recognised as the subject's leading authority, had the opportunity to extend his thoughts to the new theories of Wegener.

By the middle 1920s Holmes found himself to be almost unique among those considering the drift hypothesis in being fully cognisant of the significance of radioactive heating. He approached the various explanations given for drift with the same critical rigour he had employed over the age of the Earth. In January 1928 he presented his results at a meeting of the Glasgow Geological Society.[10] He assesses the forces called upon to tug and shove the continents and finds that 'the dominant forces involved in crustal movements must arise within the earth itself'. Since 1915 Holmes had assumed that the contribution of the radioactive atoms to the temperature of the Earth had merely slowed down its rate of cooling. To validate this view he had indulged in a circular argument whereby the decrease in concentration of radioactive atoms with depth throughout the crust is sufficient to allow it to be true. In 1925 Holmes abandoned this view – he lists a series of reasons taken from his geological investigations. Among them is that it is incompatible with continental drift: Holmes has become a convert. Like Joly he cannot believe that there is no contribution to heating beneath the crust from radioactive atoms, yet

Joly's revolutions, when the substratum becomes fiery and molten, were dismissed by Holmes as they had been by Jeffreys. Instead Holmes considered that the build-up of heat at depth led to convection in the substratum (now termed the mantle). As it took Wegener, the dreamer, to conclude that the continents were drifting, so it took Holmes, the critic of Geology, to place continental drift for the first time on a scientific foundation. No longer were the continents to be driven by forces weaker than the slightest breeze. There was a mighty engine deep in the Earth, powered by radioactivity.

The science in Holmes's account comes from his assessment of the concentration of radioactive atoms in the mantle substratum. He calculates that if in this substratum there is but 1/700th of the radioactivity of a plateau basalt, then sufficient heat is generated in 200 million years such that escape through the ocean crust would require

> one third of the whole of the ocean floors (taken at 60 km thick) to be engulfed and heated up to 1,000°C, and replaced by magma which cooled down to form new ocean floors at 300°C. A process competent to bring about this result on the scale indicated would be some form of continental drift involving the sinking of old ocean floors in front of the advancing continents and the formation of new ocean floors behind them.

He goes on to discuss the available evidence to suggest whether the viscosity of the substratum could be such as to prevent convection. He finds from the evidence of the recovery of Scandinavia after the melting of the ice-cap, that it is not. Holmes estimated the velocities of these subcrustal currents – no longer the 20 m per year required by Wegener, the rates have dropped to 'the order of 5 cms per year'.

Holmes's achievement was to have derived, or at least given the impression that he had derived, convection to satisfy certain requirements of terrestrial heating, not simply because continental drift lacked motivation. All he desired of his ideas was that they should justify 'tentative adoption as a working hypothesis'. Such hesitancy extended to the obscure publication in which he chose to publish the model. The irony of Holmes's explanation of drift was that the Englishman chose, in imitation of a German geophysicist Robert Schwinner,[11] to see the movements of the substratum by analogy with and in the language of Wegener's own science of Meteorology. 'Superimposed upon the general planetary circulation' he wrote 'there must be cyclonic and anticyclonic systems set up by the effects of regions of greater and lesser radioactivity in the overlying crust. Chief of these are the monsoon-like currents due to the distribution of continental blocks and ocean floors.'

Holmes greatest distinction was to be singled out by the mathematician Harold Jeffreys as the only geologist worthy of respect. (Both Geordies, educated in Newcastle, they had been friends since the

Great War when they both worked off Exhibition Road, London.) Since his schoolmasterly remarks against geologists made in the 1st edition of *The Earth*, Jeffreys had suffered rebuke, in particular within a 1925 review written by the Professor of Geology at Glasgow, G. W. Tyrrell:[12] 'Jeffreys has stated that the known effective forces tending to move the continents are one hundred million-million times too small! But geologists will have Huxley's mathematical mill too much in mind to pay much heed to this latest physico-mathematical *ipse dixit* upon a geological problem.'

This was a strong declaration of independence. Later that year Tyrrell was forced to issue an apology. He withdraws the association of Jeffreys with the 'physico-mathematical *ipse dixit*' but refuses to alter the sentiment, writing amidst the apology 'The remark was penned in a phase of reaction against that spirit of bright confidence in which mathematical methods are applied to geological problems by physicists and geologists alike, which has come to disaster so often in the past.'[13] Therefore, one is to infer, continental drift was acceptable to some geologists exactly because of its total disregard for the bright confidence of physico-mathematical theories. Tyrell, however, was a vociferous opponent of drift.

In the 2nd edition of *The Earth* Jeffreys quotes Tyrrell's original remarks without giving their source: 'a widespread attitude as expressed . . . by a well known geologist'. He proceeds with an angry but donnish riposte: 'Tyrrell's remarks amount to a denial of the right of geophysics to exist' and proceeds to quote Huxley's dictum on 'the mathematical mill' in full. The heart of the quotation – the centre of the metaphorical machinery – is that 'Mathematics may be compared to a mill of exquisite workmanship, which grinds you stuff of any degree of fineness; but, nevertheless, what you get out depends on what you put in; and as the grandest mill in the world will not extract wheat flour from peascods, so pages of formulae will not get a definite result out of loose data.' Jeffreys reveals that 'the mathematical mill works in both directions. We may not be sure of what we are putting in; but if what comes out is an indigestible form of cellulose, it is legitimate to infer that what was put in was not wheat'.

The context of this quotation is of great significance to the nature of the debate. Huxley's mill began to turn in 1869,[14] early in the debate between geologists and physicists as to the age of the Earth. Through invoking it, Tyrrell, and now Jeffreys in attempting to exorcise its ghostly power, are accepting that history is in danger of indulging in self-reference and even repetition. For Jeffreys this possibility had a certain truth, for culturally he was as much the heir to Kelvin and George Darwin as if he was part of a dynastic succession. A great Cambridge mathematician applying his craft to the Earth, taking an amateur interest in geology he was not concerned with lessons to be learnt from field investigations, but simply how the planet could be reconstructed through mathematics. How much he was the prisoner of this history depended on how far he was prepared to push the liberation of his own considerable intelligence. His adoption of the culture of his predecessors can be found in several details; for example he maintained Darwin's origin of the Moon out of the Pacific into the 1930s by which time it had become largely archaic, and sustained the origin of mountains through simple thermal contraction of the Earth despite all alternatives. The Moon he believed, built out of the Earth's original outer layer, was enriched in radioactive elements and was therefore much hotter than the Earth. Widespread vulcanism and degassing continued to form craters even today. The Moon was alive while the Earth was moribund. The Moon had captured all those radioactive nucleides that might otherwise upset his terrestrial heat-flow calculations. Thus Jeffreys' Earth was no more than Kelvin's or Darwin's Earth; a solid Earth of the 19th century in which radioactive atoms were strictly limited to the outer crust in order that they would not interfere with the simple physics of the interior.

That Jeffreys would always prefer a mathematical to a geological explanation is revealed in one of the appendices added to his 2nd edition. Jeffreys has supported Daly's opinion as to there being a glassy layer beneath the crust, refusing to accept geologists' beliefs that solidified material would only exist in a crystalline form and employing as a defence that 'no geologist has obtained a specimen of the matter *in situ* 20 km down' – an argument that is as spurious as Tyrrell's denials of the possibility of using mathematical analysis. Yet in his desire to kill drift, Jeffreys was even prepared to use geographical information, making a famous claim that 'the alleged fit of South America into Africa is seen on a moment's examination of a globe to be really a misfit of 15 degrees.'

After Wegener, it is Joly's turn to receive sentence. Once melting has started, Jeffreys finds 'there would be no resolidification'. 'These objections have been published previously in greater detail, and Prof. Joly's replies have consisted of a series of evasions and irrelevancies.' At the end of it all, after a short defence of the contracting Earth, Jeffreys, without quoting any of the new convecting Earth of Holmes, admits only that 'Holmes who is the only critic of the theory (of contraction) that appears to have read any work on it under twenty years old, does not base his objections on quantitative inadequacy.'

From the high vantage point of more than half a century it might seem that the debate on drift had now been reduced to whether there were sufficient radioactive atoms, and low enough viscosity in the mantle to allow convection to take place. Although this was the kernel of the argument as reduced to physical constraints, the debate was not between two well-briefed physicists; it was out in the chatter of the geological sciences. Only Jeffreys and Holmes had identified that the whole controversy could be resolved into a debate about the nature of the Earth's interior. For geologists, palaeoclimatologists, palaeo-geographers, palaeontologists and even geophysicists, their science, however little realised, was as firmly structured around the study of the Earth's surface as their own lives were committed to this same familiar territory.

The German expressionists

At the time of Wegener the German imagination was extraordinarily prolific; having identified the need for theories on the scale of the Earth, it was left to the skills of invention to arrive at some ideal, metaphorical model. Many of the theories date from the Wilhelmine period, contemporaneous with Wegener's drift, where so much of post-war Weimar Germany's intellectual culture had been initiated. The discovery of radioactivity offered a scientific demonstration of the poverty of materialist explanations. In an Earth bursting with hidden energies, fluidist theories of a molten substratum were reborn, by Ampferer[15] in 1906 and Schwinner[11] in 1920. Sluggish undertows in the substratum pushed and pulled the overlying crust into contortions worthy of the abstractionists.

Prior to Wegener's books it had been possible, as Suess, to sustain an agnosticism about whole-Earth theories, but now there was polarisation. Continental drift was itself a response to this fear of the underlying anarchy existing within Suess's great work. (Suess's anarchy had even had a political manifestation: in 1848, while still a student, he had been imprisoned for his revolutionary activities and was only released through the intervention of the head of the Imperial Geological Survey.) For the geologists there was a need to return to the golden age at the origins of Geology, before the science had become mired in the disorder of Suess and the heresy of drift. This was Geology's watershed, the beginning of the great decline. From around the time of the death of Suess, and the moment when Wegener produced his first lecture on continental displacement, as Europe entered its terrible imperial war, it was to the formerly discredited global order of Elie de Beaumont that geologists sought some certainty and inspiration.

The first agent of this reinstatement was Hans Stille, who from his inaugural lecture at Leipzig in 1913,[16] outlined a contracting Earth subject to world-wide, short-lived revolutions in which the major mountain ranges were built. While Stille's theories were still evolving, another geologist, Leopold Kober from Suess's own Vienna had begun to develop a parallel system that was turned into a classic textbook of Geotectonics *Der Bau der Erde* in 1921. Kober's Earth was also contracting; once the continents had been pliable but now they consisted mostly of stable rigid areas (the kratogens) that were separated by narrow mobile zones (the orogens). As more orogens became turned into kratogens so the kratogens would eventually encircle and weld the whole planet. Meanwhile Stille was stressing the gradual movements of the crust that took place between the revolutions; movements he termed 'undations'. After Charles Schuchert, from Yale, had attempted to name subtypes of American geosynclines, Stille was inspired[17] to define a new, more successful set of categories: 'orthogeosynclines' that gave rise to Alpinotype mountains and eu- and mio-geosynclines that differed according to the amount of volcanic activity included within them. These names were to become familiar to all Geology students up to the early 1970s; it was Stille who provided the standard orthodoxy of tectonics.

The desire to return to Geology's heroic age was no aberrant impulse. Intermittent bursts of global revolution were to become a new orthodoxy as Geology entered a phase of conservatism and reaction which was to dominate large-scale theorising. The primal war that lay at the heart of all geological conflicts – that between a disordered uniformitarianist scheme of tectonic explanation and an ordered catastrophism; the war that had been fought between Huttonians and Wernerians – was now to be fought between mobilists such as Argand and fixists such as Stille. As Argand had identified, these were attitudes of mind; in the fights to come it is not difficult to see on which side Nazi philosophy would place its natural affinities, or for that matter authoritarian Soviet Russia. The new conservative Geology offered no historical parallels for social democracy.

While Stille sustained his contracting Earth as the motor of mountain building, there were others who, since the discoveries of radioactivity, had lost faith in contractionism and therefore had need to construct still more remarkable models to confirm their fixism. In 1930, in the most reactionary of all possible anti-mobilist stances, a German geologist, Erich Haarmann, denied all horizontal movements, preferring to implicate ideas that had been suggested in the early 19th century but that were now to be turned into a universal principle.[18] Horizontal movement was all the result of the gravitational collapse of layers of rock, sliding off the sides of uplifted crustal blocks. In *Der Oszillations-theorie*, these uplifts were inelegantly termed 'geotumours'; they followed after episodes of crustal depression that provided the geosynclines. This was the coldest, most austere and rigid of all tectonic theories; a world-view for the 1930s. At the 1931 Berlin meeting of the German Geological Society Haarmann's views were roundly dismissed, but yet he had obtained a number of influential converts to his vertical tectonics (such as van Bemmelen and Ramberg) who were to pursue studies that would continue to 'prove' that all mountain ranges contained an illusion of horizontal motion; that nappes were deceptive landslides. Haarmann's ideas were also to provide the foundation of a new state-approved geological orthodoxy in an eastern neighbour.

For all these geological whole-Earth theories, it was less important that they should be physically plausible and more that they satisfied certain rigorous and traditional demands implicit in the science of Geology. The central geological argument against continental drift, was that it furthered Suess's heresy through failing to satisfy the demand for universal diastrophism; world-wide revolutions. Geology had become a political agent. Only in Britain was the pluralist desire to satisfy uniformitarian precepts so strong that geologists did not even make the attempt to construct elaborate whole-Earth models.

This tension between the uniformitarian ideal and the desire for past global revolutions existed deep within the science, for without global events, world-wide geological correlation became impossible. The periods, Jurassic, Triassic, Permian had to be more than the parochial reigns of English monarchs; they had to be global. Although the means of detailed correlation, the study of fossils, had come from William Smith, the names of the geological periods based on the appearance of the rock (Cretaceous, Carboniferous) went back to Werner and his belief that their boundaries were sudden and world-wide. Diastrophism was all that gave the science its universal language. Elie de Beaumont had attempted to find the physical cause of such universal revolutions but even without such explanations, these events satisfactorily provided the interruption in the continuity of geological time, like the chimes on a clock.

In Germany, the melting pot of new theories, it was inevitable that the opposition to the contracting Earth should produce its antithesis. These theories took Wegener's drifting continents and supplied an underlying inflation as first proposed by Mantovani. Such an adaptation is unremarkable, because on close scrutiny Wegener's theory is implicitly in accord with an expanding Earth. Wegener's drift was not considered as a process but as the fragment-ation of a single continent of unknown origin: Pangaea. Those oceans that concern Wegener are only those that have come into being since that disintegration began: the Indian and Atlantic. The con-tinents and oceans have evolved out of a single historical land mass. Topologically and mentally Wegener's planet has expanded; and from the flight of his beloved Greenland, in modern times, at increasing speeds. Thus expanding Earth theories were forever to be found emerging independently from those scientists who studied Wegener. In 1927 Lindemann wrote a book claiming that the tensional Earth was being blown up by gases in the Earth's interior.[19] In 1933 Hilgenbirg, blaming interior aether sinks, published an account dedicated to the memory of Wegener, in which he demonstrated on a series of papier mâché spheres that a reconstruction on a smaller globe eliminated all the oceans allowing the modern continents to cover the Earth.[20] This was a perfect theory for the geological unconscious; as Geology was the science only of the land, and as almost nothing was known of the ocean floor, what better than to show that there had once been no ocean basins; and therefore that Geology already investigated all that was worthy of consideration? Hilgenbirg even dusted down his theory during the continental drift revival of the late 1960s.[21]

Immediately after the Great War there was in Germany a flood of pseudo-scientific pamphlets; of myths and magic and of science fiction. There was a clamour for a 'spiritual reorientation' in the face of the tragedy of defeat and the breakdown of government. 'Sensations and imaginative fantasies . . . would help to conquer gross 'materialism' and its literary counterpart, realism.'[22] Wegener's drift could not be assimilated as myth, or not at least in the manner of the most outrageous and successful of the post-war cosmologies: the *'Welt Eis Lehre'*, the creation of an Austrian technician Hans Horbiger.[23] A large organisation was set up in Germany to proselytise Horbiger's cosmic history, and an alliance was forged with the Nazi Party. In 1925 Horbiger wrote to all scholars in Germany and Austria stating that 'You now have to choose to be with us or against us'. Once the Nazis were in power the theory was given national approval as being authentic history and Adolf Hitler proposed that a commemorative statue of Horbiger should be erected in Linz, Austria, alongside those of Ptolemy and Copernicus.

The *'Welt Eis Lehre'* was a mixture of problems and explanations; much of them gleaned from Geology. Thus the problem of the Ice Age was to be solved by the crashing onto the earth of an ice-satellite. There had in the past been primary, secondary and tertiary ice-moons, each of which had in turn collided with the planet to bring to an end the Wernerian geognostic epochs. Such collisions occurred when the peoples of the Earth indulged in moral turpitude and racial impurity. The most recent collision had brought about the Ice Age that had marked the end of Atlantis-Thule, the original home of the Teutons. As in the cosmologies of Taylor and Chamberlin, it was the arrival of satellites to orbit and collide with the Earth that provided the great trigger of terrestrial activity. The Quaternary moon hovered uncertainly over the heads of the German people, already engaged in its inevitable descent to the Earth to provide the next great wave of purification.

Horbiger's cosmology offers some insight into Wegener's insistence that Europe and America had been joined to Greenland right into the most recent geological period. Ice lay at the heart of both Horbiger's and Wegener's histories. If its position was irrational then it was also integral; Wegener's continental drift could not be sustained without this ingredient. The need to explain the Ice Age; the shadows, however faintly perceived, of some lost North Atlantic continent lie within Wegener's explanation.

America: the triumph of fixism

The age of ice and steel that the Nazis inflicted on the German peoples effectively concluded any significant contribution that could come from this country. America willingly received the refugees, among whom was the renowned German Jewish seismologist Beno Gutenberg who proposed his *'Fliesstheorie'* of continental spreading to explain Wegener's palaeoclimatic observations.[24] The single continent left after the loss of the Moon, had spread from the South Pole, under the influence of the Earth's rotation, to provide the higher latitudes of Eurasia and North America.

In America the absence of anyone who ever considered themselves a straightforward supporter of Wegener ensured that the dark ages of drift were most thoroughly benighted. After the death of Wegener, Taylor, fed up with being manacled to an unfashionable ghost, plucked up sufficient courage to plead that 'since the views of the two authors of the Taylor–Wegener hypothesis differ in several respects, and the first named author's paper preceded the earliest of Wegener's published work, it seems best to discontinue the hyphenated relation'.[25] Taylor died at the site of his birth, Fort Wayne, Indiana, in 1938. His petition for divorce was granted by history; the hyphen disappeared, taking with it Taylor's own name.

For Taylor had all his life been an insubstantial, ghostly presence on the American scene. While frequently invited to discuss his theory, there was little, if any, criticism directed against it. Taylor was tolerated and ignored as an American eccentric; no one ever suggested that he was plotting something radical or disturbing, likely to reconstruct Geology. Having determined that the Moon arrived in some Cretaceous June, his theory never changed, never absorbed more physical or geological data, but remained like a saga told by an old man, a familiar, even popular, tale to be listened to and then dismissed. The tolerance of Taylor is but one facet of the American hostility to Wegener; Taylor was almost encouraged as the home-grown drifter – proof that America did not persecute such individuals simply because of their beliefs.

In common with their German colleagues, American geologists had come to the realisation that it was up to them to fill the previously undiscovered territory between Astronomy and Geology. The lead had already been taken by Chamberlin – to make Geology command a greater cosmological empire; and as a reflection of the new era, Geology textbooks after Stille and Kober no longer considered themselves to be of the type 'A student's Lyell' but instead offered overviews of Geology complete with a more tentative cosmogonic or visionary whole-Earth theory. Not all writers were in the enviable position of Chamberlin of being the editor and founder of the journal in which such theories could gain publication. On Chamberlin's death it was his obituarist, Bailey Willis, emeritus Professor of Geology at Stanford, who chose to provide the traditional geological overview. In February 1928 he wrote in *Scientific American*, 'Growing mountains'.[26] After a decade of turmoil in whole-Earth theories he wishes to return the general public to old-fashioned geological commonsense. After discussing the seductive charms of mobilism he states 'Others, of whom I am one, incline to the theory of a good solid Earth.' In searching for a metaphor for the mobilists, Willis ironically stumbles upon an image that is the most revealing of Geology's spatial scope: 'We may admire the intrepidity of these intellectual aviators, but we fly with them only at the risk of a crash.' This fear of flying was to continue to keep Geology grounded.

The new German school of global Geology also had its younger imitators. The greatest of the American textbook theorists of this period was Walter H. Bucher, whose whole-Earth model first appeared in 1924. While disavowing drift, his book has a self-conscious allusion to *The origin of the continents and oceans*. And as proof of the cultural roots of these new Earth models, Bucher had been educated in Germany, completing his Ph.D. in 1911, the same year as Gutenberg.

Walter Herman Bucher provided a strong, coherent American Geology which offered a fully articulated alternative to mobilist theories. Born in Akron, Ohio, on the completion of his German education he entered the Geology Department at the University of Cincinnati where he was to remain for 27 years. A series of papers culminated in a book *The deformation of the Earth's crust*, published in 1933. From geological facts, Bucher intends to derive inductively 'a hypothetical picture of the mechanics of diastrophism'. These geological facts are to be listed as 'laws', for 'not until we have succeeded in the formulation of a body of specific laws, can we expect to arrive at an intrinsically satisfactory solution of all problems of geotectonics.' What is this sleight of hand? Bucher is attempting to show that Geology is a science as rigorous as any other, and that as there are laws of mechanics and thermodynamics, so there will be Bucher's laws of Geology. The heavy germanic style of the book, paragraphs headed with bold-type 'laws', followed by italicised 'opinions', resembles a logical positivist philosophy text – which was the intention. For Bucher believed that drift was the erroneous result of subjectivism; he even discusses the philosopher Vaihinger's analysis of the innate logical fallacy of suppressing the 'inconvenient little words "as if"', in those writings of drifters who believed that they could somewhere observe evidence of motion. From materialism to logical positivism – this is the philosophical shift that Bucher is attempting to engineer in order to rebuild Geology for the 20th century.

'Furrows' are the great linear trenches down which sediment pours, ultimately to be turned into new mountain ranges or 'welts'. To find the significance of the global pattern of welts and furrows, Bucher resorts to experiments with a variety of balls. He tries deflated rubber balloons coated with paint, wax, gum arabic and gelatin; rubber balls, inflated with air, covered with a mixture of paraffin and vaseline and deflated slowly under water; paraffin-coated rubber sponge balls; and the thin glass spheres used as ornaments on Christmas trees, subject to high external pressures. Among all these toys there seemed little to compare with the Earth. However, undeterred, Bucher had continued with the Christmas tree decorations. He filled the glass spheres with water and froze them. Success! The yule-tide science had shattered the sphere and modelled the Earth. The pattern of geosynclines, of linear furrows on the planet, was caused by phases of expansion. Yet the mountains themselves revealed that these linear deeps had subsequently been squeezed. Bucher had arrived at his cosmology: 'The essence of the hypothesis is that epochs of crustal tension have alternated with epochs of crustal compression.' Bucher has already seen the portentous philosophical implications: 'It substitutes a dualism for the monistic views which consider one continuous process as the cause of all crustal deformation.'

Despite its logical positivist structure, Bucher's work contains all the most fundamental ingredients of the geological philosophy of Elie de Beaumont. Orogenic episodes are world-wide revolutions related to a single underlying global mechanism. He quotes from Stille's list of orogenic epochs and lists 24 such episodes since the Ordovician geological period. Naturally those principles that dictated the deformation of the continents, where geologists had access, were identical to those hidden processes that controlled the greater morphology of the ocean floors. The Earth's interior was homogeneous and unintrusive so that the crust of the Earth, the province of Geology, could itself reveal the whole story. The cause of the dualism was presented in the third of his 'hypothetical conclusions . . . of fundamental importance . . . alternating swelling and shrinking of subcrustal material . . . [from] fluctuations in the heat content of the subcrustal body of the Earth'. The explanation appears to be taken second-hand from Joly. Bucher ventures no further.

Bucher's work fitted in to the traditions of Stille and Kober and ensured that Geology could demonstrate a concern for an intellectual tectonic overview. Yet the bridge he offered between the scope of whole-Earth theories and that of the rocks was illusory; Bucher's Christmas tree decoration whole-Earth was considerably smaller than the mountains in which its history was contained.

Bucher was extremely influential; as the Chairman of the Geological Department at Columbia he received many honours for his insights into geotectonics and in 1954 was elected President of the Geological Society of America. He died as plate tectonics was born in 1965. His textbook was reprinted in 1957, 1964 and 1968. For these subsequent editions the book remained in the timeless fashion of fixism as defined by Argand, unaltered, unalterable. In a new preface, Bucher describes his laws as providing 'a touchstone for every man's opinion concerning crustal deformation', but already in the 1957 preface he reveals that 'the more we learn about the incidence in time of tensile and compressive deformation in the crust, the more it becomes plain that these two types of deformation cannot represent a world-wide rhythm.' That this recantation has not required the writer to alter the text is remarkable. The ordered pattern of revolutions has all gone; yet the book has not, cannot be, rewritten because every page was imbued with the certainty that Geology *could* derive an ordered whole-Earth theory. Without the greater structure, Geology is once again parochial.

Just as Elie de Beaumont's universal geometry had finally collapsed under the need for continued subdivision, so universal correlation was breaking down as the time periods of orogenic episodes threatened to multiply into a continuum. The bright confidence in the existence of an ordered global geological history had now broken into entropy. Without such order, Geology was no more than disconnected observations. Bucher's casual apology is a symptom of the loss of Geology's nerve. No new theories of the Earth were developed by geologists through the 1950s. After a phase of buoyant reaction in the 1920s and 1930s the science had run out of ambition and become withdrawn.

At this lacuna there burst upon the scene an extraordinary charismatic wizard who succeeded in inciting the established sciences into a frenzy of outrage. Immanuel Velikovsky, a Russian Jew born in 1895, had been a medical student in France, Edinburgh and Moscow, before moving to Palestine to begin an apprenticeship in psychoanalysis, meeting and corresponding with Freud. In 1939 he travelled with his wife to New York to begin research on Freud's three heroes of the ancient world: Oedipus, Akhnaton and Moses. Velikovsky began to discover accounts of strange and terrible catastrophes and chose to widen his search into the ancient writings and myths of all nations and civilisations.

Having decided, a priori, in the manner of Stille and Elie de Beaumont, that such catastrophes were global events, outside the realms of ordinary uniformitarian processes, he followed the same geological reasoning in synchronising these events, even though this might require distorting the historical timescale. It was an inevitable development; Wernerian geognosy that had been based on the biblical account of Creation had been accommodated by Geology as Cuvier's catastrophism in which there was a repetition of disaster. Now catastrophism was to be reapplied to the Bible. Geology had sown the wind, now it would reap the tornado, cutting a slow swathe of disorder across the intellectual landscape. Like Donnelly, Velikovsky was concerned to integrate scientific findings with the primary written history. Like the myths that he chose to interpret, Velikovsky wished to find a single thread or theme that would conjoin all the separate cataclysms. He found it in that libido of the spheres: the planet Venus. His work became a psychoanalysis of the solar system.

An American, Howard B. Baker had first proposed[27] in 1911 that a close osculation with Venus had provoked the Moon to burst from the Earth. Velikovsky sought explanations for more recent biblical catastrophes. Venus had emerged parthogenetically from Jupiter. During its crazy flight across the orbits of the planets its comet's tail had brushed with Earth to cause the rain of fire; the Earth had tumbled in space to make the Sun appear to stand still, and Venus had collided with Mars, which so perturbed had also intermittently approached the Earth. The tumbling Earth explained the confused pattern of past climates recorded in the rocks; electric discharges between the planets had shifted the magnetic field; vast floods had swept across the land carrying erratic boulders, sweeping the ocean floors free of all but thin layers of sediment; and the rocks themselves had thrust up to form mountains.

Velikovsky's theories first appeared in 1946 in the *New York Herald Tribune*, where the science editor termed it 'a magnificent piece of scholarly research'. In 1950 the first book *Worlds in collision* was published by Macmillan, to great popular acclaim. Scientists were aghast. The Yale Professor of Geology, Chester Longwell, wrote a review for the American *Journal of Science* in which his outrage was directed at the publishers, who had listed the volume in their catalogue under 'Science'. Yet Velikovsky had touched a raw cultural nerve. Scientists communicated through written texts; the scientific paper was the established method for a scientist to claim priority in an idea. Therefore the written word was central to the transmission of scientific data. However, Geology had flourished through offering a more accurate history of the world than that to be found in written texts. Yet the written word of the Bible still formed the central backbone of the Christian nations. Here was a paradox. The decision to select some written works as truthful and others as erroneous is a subtle one. To a psychoanalyst they are all equally revealing. The truth of the Old Testament demanded a new cosmology of repeated catastrophes. This Velikovsky had provided in the full knowledge that the Bible was the truthful history to which all other scientific ideas must be subsidiary.

Only one geologist came forward to offer Velikovsky a hearing. In 1952 Velikovsky had been researching his second book *Ages in upheaval* at Princeton when he encountered the head of the Geology Department, Harry Hess. In 1956 Hess agreed to pass on some of Velikovsky's proposals for the planning of research projects in the International Geophysical Year. In 1965 Hess asked Velikovsky to lecture to his students on his wealth of knowledge of history, myth and catastrophism, as a challenge to their conventions, and soon other campuses had emulated the invitation.

It is no coincidence that Harry Hess, the only geologist to offer the crazy 'Old Testament' catastrophist a hearing, was also to be the only senior geologist in North America who would be involved in the revolution that was soon to overtake the science. To support Velikovsky was to demonstrate a certain scepticism about the function of the geological orthodoxy. This was particularly brave. It has been claimed that active support of drift by a staff member during this period would prevent the possibility of academic employment. As there had been no single US native supporter of Wegener, *émigrés* provided an important source of mobilist sympathies. The most influential of these was probably John Verhoogen, a graduate from Brussels and Liège who began promoting drift (powered by mantle-convection) in seminars at Berkeley where he taught from the late 1940s.

In the new post-war prosperity the rebellion of students against the wisdom of their teachers became more commonplace. Continental drift had considerable appeal to the irresponsible imagination. On 28–29 December 1949 a symposium on the South Atlantic was held in New York by the Society for the Study of Evolution. Simpson spoke vehemently against drift and Bucher and Maurice Ewing (see Chapter 6) summoned the geological and geophysical evidence for permanent continents. One speaker, Dr A. S. Romer, indicated the political overtones of his belief in past connections across the South Atlantic, remarking that 'to my embarrassment . . . in such a "leftist" position I am disturbed . . . by the company of bridge builders, radical continent shifters and Gondwanaland collectivists which this may entail.'[28] In 1951 Professor Lester King, from South Africa was enticed into a student debate at Columbia to face the formidable Walter Bucher. The result, by repute, was a considerable victory among student sympathies for drift. It was up to geologists to convince their students that such theories, like communism, although alluring, were no more than indulgent fantasies. Bucher himself termed the supporters of drift 'Wegenerians'; the term had great resonance with another term of disparagement employed at the origins of Geology: Wernerian.

The far-flung outposts of drift

The foreign geological culture from which Lester King had sprung had sustained consistent support for drift as the result of one man's efforts. Alexander Logie Du Toit, the first in the English-speaking world to have written in praise of Wegener,[29] was also drift's most ardent champion. Born in 1878 near Capetown, he had studied Mining engineering in Glasgow and Geology at the Royal College of Science, London, before returning after the Boer War in 1903, to join the Geological Commission of the Cape of Good Hope. From 1914 he began a series of visits to every fragment of Suess's Gondwanaland of which South Africa formed the heart.

**James Alexander Logie Du Toit,
1878–1948**
South African geologist

106

When Reginald Daly arrived in South Africa to learn for himself the significance of Gondwanaland, a friendship was sealed that led to Daly assisting Du Toit in gaining a grant from the Carnegie Institute for a five month visit to the eastern nations of South America. Du Toit's account of the similarities between the geology of the two continents[30] was enthusiastically received by Wegener. Daly considered Du Toit to be 'the world's greatest field geologist'.[31]

In 1937 Du Toit wrote *Our wandering continents: an hypothesis of continental drift*, a book that was dedicated to the memory of Alfred Wegener 'for his distinguished services in connection with the geological interpretation of OUR EARTH'. In a synthesis of all mobilism, Du Toit undertook to pull Wegener's theories into a more coherent shape and to 'geologicise' them, offering a cause in mantle convection and subcrustal melting. In place of the fragmentation of a single continent of Pangaea, continental drift was shown to be a uniformitarian principle that could explain past mountain ranges.

While Holmes had seen so acutely that the nature of the continental drift debate was, for a physicist, resolvable into a dispute as to the viscosity and radioactive content of the substratum, Du Toit resolved the dispute for a geologist into the battle between horizontal and vertical forces. He quotes from the American Association of Petroleum Geologists symposium 'In pointing out the difficulties introduced by isostasy, Schuchert voices the pathetic appeal, "Geologists must find a way to sink land bridges." Why? Save to extricate orthodoxy from an impasse! For the words "sink land bridges" we might with at least equal fairness substitute "move the continents".' As Wegener had recognised, isostasy was the most fundamental of all the 'laws' to apply to the Earth. Yet for most geologists it was patently simpler to sink a land bridge 4 km than to move a continent 4000 km.

Unlike Holmes, who seems to have arrived at a support of drift through a long and reasoned process of discovery, for Du Toit drift was a self-evident truth, a truth that had dawned on him on first hearing of Wegener's work and that burned fiercely through all his endeavours. A geological theory had to be demonstrated to be accepted; there was no way available, apart from Wegener's now discredited measurements of the flight of Greenland, to prove drift. In his book Du Toit combined his conviction with his frustration at being unable to offer a scientific proof. Frustration also haunted his literary style. Like Wegener, who had been accused before him of playing the advocate, Du Toit had to bellow his enthusiasm. With all the elegance of a long journey on a rock-strewn dirt road in a car with no suspension, Du Toit describes the experience of apostasy:

The geologist will have to leave behind him – perhaps reluctantly – the dumbfounding spectacle of the present continental masses, firmly anchored to a plastic foundation yet remaining fixed in space; set thousands of kilometres apart, it may be, yet behaving in almost identical fashion from epoch to epoch and stage to stage like soldiers at drill; widely stretched in some quarters at various times and astoundingly compressed in others, yet retaining their general shapes, positions and orientations; remote from one another throughout history, yet showing in their fossil remains common or allied forms of terrestrial life; possessed during certain epochs of climates that may have ranged from glacial to torrid or pluvial to arid, though contrary to meteorological principles when their existing geographic positions are considered – to mention but a few such paradoxes!' – intake of breath.

Du Toit was the first to have thought through the implications of continental drift for Geology. 'Looking back' he begins his book 'dispassionately into the history of Geology it is interesting to observe how deeply conservatism has become entrenched'. To understand why the 'New Geology' 'has apparently found so few whole-hearted supporters' Du Toit finds firstly that 'it cuts at the basis of customary geological interpretations'; an 'objection that is largely a psychological one'. To be overcome 'this would involve the rewriting of our numerous text-books, not only of Geology, but Palaeogeography, Palaeoclimatology and Geophysics.' Du Toit even justifies his enthusiasm: 'a world without some form of crustal drifting would appear . . . as unreal as one lacking in biological evolution.'

The balance of power between the sciences is still that of the 1920s: 'To the geophysicist pre-eminently must be left the search for and evaluation of an adequate force for such crustal movement as is indicated by the full weight of geological data and inference.' Yet there remains a poignant tension between Du Toit's insistence on making drift compatible with the philosophy of Geology and at the same time recognising that the proof of continental drift would create a different science, termed 'New Geology'. This proof, he has identified, awaits the discovery of the force providing crustal movement; a discovery that must emerge from Geophysics. Implicitly, this 'New Geology' will contain an altered balance of power between geophysicists and geologists.

The publication of Du Toit's book had little impact except to arouse the American critics into new heights of ferocity. In 1943 an American vertebrate palaeontologist, George Gaylord Simpson, launched the second wave of the tirade against drift by reviewing in the *American Journal of Science* all the evidence of the distribution of both fossil and living mammals.[32] Stung by the repeated claims of drifters for corroboration he responded 'It must be almost unique in scientific history for a group of students, admittedly without special competence in a given field, thus to reject the all but unanimous verdict of those who do have such competence'. Land bridges, now termed by Simpson with analogy to Central America 'isthmian links' could explain all. He was not extravagant in his demands for such corridors, and thereby felt justified in censoring the excesses of others. Even Du Toit himself had needed to build occasional bridges to explain details of the fossil record.

Simpson explained the derivation of one of these bridges. In January 1919 the French geologist Leonce Joléaud[33] believed from some published identification (subsequently revealed to be a mis-identification) of fossil teeth, that in the Miocene geological period there had been a land bridge between Florida and Europe. The teeth were of a genus of primitive horses known as *Hipparion*. The bridge once postulated began to grow and soon was claimed to have persisted through much of the Tertiary. In 1924 Joléaud became a convert to drift, yet still required sporadic interchange of species, which he expedited through allowing the continents to come and go '*avec un mouvement en accordéon*'. Joléaud's original *Hipparion* bridge, now abandoned in favour of the new music of drift, was fortified in the presidential address given to the Geological Society of London in 1929 by J. W. Gregory, from whence it was claimed by Du Toit as one of the land bridges for which the evidence was incontrovertible.

As Simpson remarked, Du Toit neither admitted that the biological evidence for drift was completely convincing nor that it was possible to do without the occasional supplementary land bridge. In which case, why bother to invoke the monstrous displacement of continents? Simpson's alternatives were economical and therefore convincing. The faunal similarities of Gondwanaland had arisen because a common assemblage of species had fled north on to the separated equatorial continents at the coming of an ice age; the status of the species on Madagascar was the product of a 'sweepstake' – the chance arrival of animals swimming or rafting from the mainland.

In March 1944 Du Toit replied,[34] claiming that limiting the discussion to the seldom fossilised and relatively recent Mammalia was a serious misrepresentation of the great wealth of supporting evidence from alternative and older fossil species. Simpson had by now gone off to fight in another global war, and Chester Longwell chose to reply on behalf of his colleague.[35] The concentration on mammals relates to Simpson's own specialty; other experts could speak up in turn to show how their own discipline contradicted drift. Longwell again brought up Wegener's correlation of the ice-front moraines between Europe and America, and then more damningly listed some of the new radio distance measurements made across the Atlantic. From 1913 to 1927 one report showed an increase of 0.32 m per year between Paris and Washington; another a decrease of 0.7 m per year between London and Washington. Du Toit had attempted to save drift, by restricting Wegener's speedy flight of Greenland from Europe to a snail's pace of only a few centimetres each year, but Longwell, in the manner of the Americans, preferred to sharpen his criticisms on Wegener's more outrageous demands.

In 1944 an elderly Bailey Willis attempted to have the final word. In an article titled 'Continental drift, ein Märchen'[36] (a fairytale) he demanded with some small-town book-burning zeal that drift be outlawed 'since further discussion of it merely encumbers the literature and befogs the minds of fellow students. . . . Scientists who are not geologists cannot be expected to know that the geology upon which protagonists of the Theory rest assumptions is as antiquated as pre-Curie physics'.

Du Toit died in 1948 and so the visit by the South African Lester King to America in 1951 provided the next round in the argument after Willis's appeal for disqualification. If 1951 marks a turning point in the American response, where hostility had reached its zenith, and where there were the first glimmerings of a new sympathy, then the call to discuss the present state of the drift hypothesis at the 1950 meeting of the British Association[37] also marks a post-war desire to reassess this perennial controversy.

The debate is the same, only some of the antagonists have changed – new actors for old roles, with some performers there at the beginning still declaiming. As ever there was Professor Harold Jeffreys: 'this is the fourth time that I have taken part in a public discussion of this theory.' Petulance has given way to an avuncular tolerance. He is talking to a group of children: 'We have all learnt at school that fluids settle down with level surfaces, and solids do not.' He details the various reasons as to why the Earth's mantle possesses strength, is a solid, does not flow and ends by attempting to blow a whistle to halt the endless playtime squabbling: 'I think that the time has come when a general account of the facts of the past distributions of fauna and flora should be produced without regard to a particular theory'. But still the dispute continued along familiar lines. Geophysics, orthodox Geology and Geomorphology were opposed to drift; Palaeobotany, Palaeozoology and Southern Hemisphere stratigraphy were in favour. This was deduction versus intuition; the study of causality against the study of relationships.

Of all nations, none contained geologists more instinctively and enthusiastically supportive of drift than the Netherlands. Van Waterschoot van der Gracht had had the zeal of a missionary amidst the hostile fixism of America. The British, the French, the Germans and the East Coast Americans were encouraged to learn Geology within their respective nation's frontiers, and thus Geology in all its dead stability. The Dutch, having no Geology apart from that buried beneath piles of silt and sand, naturally roamed to where Geology was at its most exciting. They voyaged to the mountains of Europe, where they saw the evidence for horizontal movements that had so inspired Suess and Argand, or even more critically to the Dutch East Indies.

The lessons learnt in the colonies seemed difficult to forget. The problem in convincing geologists of the truth of drift was always that it could not be seen. If there is one location on Earth where drifting land masses become most self-evident then it must be these islands that lie at the crux of Wallace's zoological paradox. One has only to listen to the Dutch geologists themselves. At the 1926 symposium[38] van Waterschoot van der Gracht spoke of the experience of these places as though it was the revelation that had effected his conversion:

The extremely active manner in which New Guinea has . . . pressed . . . into the great Soenda festoons, is so obvious that the *fact* cannot be denied by any one who is really familiar with these regions, regardless of its causal explanation. This is the reason why the Dutch geologists (Molengraaff, Brouwer, Wing Easton) who worked in the East Indies are invariably favourably inclined to Wegener's hypothesis. I have also visited this area: the evidence is indeed striking. Without knowing why, we *see* that New Guinea drifts violently to the north.

The effect that this region had upon its geologists merely required the discovery of Wegener's works to be comprehended. However, as van Waterschoot van der Gracht found to his cost, converts to drift could not be made just by recounting the experience of Borneo and the Celebes. It was necessary for the novitiate to travel and work in the jungles. The most influential of these East Indies converts was not to be a Dutchman but a Tasmanian.

The new territories

At the British Association meeting in 1950, Jeffreys' masterly admonishments, the enthusiasms of the palaeogeographers, the disconcern of the geologists, all displayed how little the form of the debate had altered in 30 years. To call it a debate is to suggest a fluidity of position, a flux of argument; instead this was trench warfare long after any enthusiasm for battle had expired. To affect in any way this cynical stalemate required a naïvety that only young researchers could supply, above all with new forms of quantitative, physical research that could hope to transform the argument from one of impressions, beliefs, contentions, assumptions, and to provide a new weaponry of cold scientific facts. This arsenal began to be assembled soon after the 1950 meeting, the first significant initiative since 1921. The new science was the study of Earth magnetism.

Just as the compass needle had for centuries pointed a fixed course for navigators, explorers and all who were lost in unknown territories, so the magnetic needle imprisoned in the rocks could hope to track the paths of continents in their journey through featureless prehistory. Magnetism had provided the first science of the Earth. When Gilbert published his *De magnete* in 1600, containing the results of experiments on the magnetic properties of a ball of lodestone, he showed that the Earth was in effect a great magnet.

By the beginning of the 20th century the Earth's rotation was known to be somehow responsible for its magnetism, for the magnetic field is too quixotic and the Earth too hot and too old to be the equivalent of a giant boulder of lodestone. As the north and south magnetic poles lie only a few degrees displaced from the axis of rotation, the dip of a magnet allowed to swivel around a horizontal axis is dependent on

111

latitude. At the North Pole the dip is close to vertical; at the Equator close to horizontal. In 1926 a French geophysicist, P. L. Mercanton, wrote that from the direction of the magnetic field recorded in lavas as they cooled below the Curie point (the temperature below which appropriate minerals can be magnetised) one could hope to show whether the continents had drifted.[39] The magnetic dip could measure past latitude and the compass-bearing azimuth of the north magnetic pole would reveal continental rotation. However, in 1926 instruments and techniques were not available for the determination of such sensitive parameters.

A new interest in magnetism was reawakened at the beginning of the Second World War after the Germans dropped magnetic mines along the east coast of England. The scientists who were put in charge of mine-sweeping and other countermeasures included Edward Bullard and Patrick (later Lord) Blackett. They argued fiercely about the origin of the Earth's magnetic field. Bullard believed that the magnetic field was generated in the Earth's outer core and was to develop his theory of a 'self-exciting dynamo' (based on a proposal made in 1946 by Walter Elsasser,[40] a German geophysicist who had moved to America in the 1930s). However, Blackett preferred an alternative explanation, one that could evade the problem of how a body as hot as the Earth's core, or even the Sun, could generate a magnetic field. He proposed to test a model whereby it was the rotation of a massive body that produced the field; a model that could hope to lend support to Einstein's cherished belief that there existed a unified theory of gravity and electromagnetism.

Blackett had originally been a naval officer in the First World War and became the father of Operational Research during the Second. His great admiration for the Soviet Union was to prevent him gaining a visa to America during the McCarthy period and also to remove him

Lord Patrick Blackett, 1897–1974
British physicist

from the British atomic bomb project at the end of the war. However, Blackett's wartime role had gained him many friends in Government and enabled him in 1952 to borrow 38½ lb (17½ kg) of pure gold from the Bank of England to be constructed into a cylinder in his Imperial College laboratory. In order to be able to detect the anticipated faint magnetic field generated by the rotating gold he required an instrument more sensitive than any yet available and so built one of his own design: an 'astatic magnetometer'. The experiment was unsuccessful, therefore lending more credibility to Bullard's theory (further supported by measurements undertaken by one of Blackett's research students, Keith Runcorn, who found the appropriate increase in the magnetic field down a Lancashire coal-mine)[41]. Blackett returned the gold, and was now left with a highly sensitive magnetometer in search of a problem.

During the 1940s various attempts had been made to map the shifts in the direction of the Earth's magnetism recorded in recent sediments.[42] Blackett's instrument was capable of detecting not only the strength but also the direction of very small magnetic fields – as possessed by rocks. Around this instrument there soon grew a small research school of rock magnetism. The presence of magnetic minerals (such as magnetite) allowed many lavas and some sediments to store a remanent magnetism, imprinted at the time of the rock's formation, unless the rock had suffered lightning strike, or too sound a beating with hammers (that Jeffreys was later to claim had affected all readings). The specimen had to be oriented before collection and extricated gently.

Some of the initial specimens came from England; in 1954 the group began to publish their extraordinary results.[43] On the assumption of an orientation of the magnetic field close to that of the Earth's axis of rotation, England, the foundation block of permanence, in the Triassic period (200 million years ago) had been twisted around by more than 30 degrees and lodged somewhere close to the equator. By 1956 rocks from India had revealed that the subcontinent had switched allegiances from the Southern to the Northern Hemisphere during the most recent geological periods.[44] Blackett had been quick to realise the significance of these discoveries, claiming in 1954 that 'the result of this work will, in the next decade, effectively settle the main facts of land movements, and in so doing will have a profound effect on geophysical studies of the earth's crust.'[45]

The magnetists 80 km to the north of London, at the Cambridge Department of Geodesy and Geophysics, preferred a different interpretation. Polar wandering had been promoted ever since the mid-19th-century discovery of the Ice Age: 'We may not merely admit, but assert as highly probable' wrote Kelvin in 1876, 'that the axis of . . . rotation . . . may have been in ancient times very far from (its) present geographical position'.[46] Kelvin's distaste for evolution became entangled through his supervision of the work of Darwin's son, George. The young research physicist had chosen as the topic for his first scientific paper: 'the fixity or mobility of the earth's axis of rotation',[47] and had concluded that the poles could wander if the Earth was plastic as he, for one, thought that it was. The condemnation received from the physicists for contemplating such a dangerous notion delighted Charles Darwin who wrote to his son 'Hurrah for the bowels of the earth and their viscosity and for the moon and for the Heavenly bodies . . . and for my son George . . .'.[48]

In 1911 an electrical engineer, Hugh Auchincloss Brown, heard of frozen mammoths found 'with buttercups still clenched between their teeth'. The accumulation of ice at the poles, he reasoned, may periodically (every 8000 years) upset the spinning planet, sending it tumbling 'like an overloaded canoe'. Brown became a great publicist[49], petitioning Congress to warn that the next such somersault was imminent, at which time the flood of waters across the globe would leave the Eskimos to inherit the Earth. In 1946, at the birth of the UN, he formed 'The Global Stabilisation Organisation' which sought to use whatever means available, including atomic bombs, to limit the dangerous build-up of ice on the top-heavy Antarctic ice sheet. During the 1950s a slower version of Brown's tumbling Earth became championed by a New England history teacher, Charles Hapgood, and an elderly engineer familiar with gyroscopes, James H. Campbell.[50] In 1955 they presented their ideas to a specialist panel at the American Museum of Natural History in New York, where the chairman Bucher expressed doubt that the Antarctic ice-cap was increasing in thickness sufficiently to pose any threat. Most scientists had anyway already dismissed these scare stories, pointing to the Earth's Equatorial bulge that keeps the axis approximately fixed.

However, 1955 was *the* year of polar wandering theories. In 1955 the polymathic English astronomer, Tommy Gold, then at the Royal Observatory at Greenwich, wrote in *Nature*, echoing George Darwin, that the Equatorial bulge provides no such stabilisation if the Earth's interior is plastic, for then the bulge too can wander.[51] The rate at which the pole can move is determined by the rate at which the substratum can flow. Shifts in the distribution of mass on the Earth's surface could come from several causes. Walter Munk, at Scripps Institute of Oceanography in California discussed the shift caused by the formation of asymmetric ice-caps in the middle latitudes.[52] If the rotational pole could wander, then it took the magnetic pole with it, for the study of the shifting magnetism preserved in recent clays from New England had shown that although the magnetic pole wandered from year to year it never roamed too far from the Earth's axis. In 1955 Keith Runcorn, now moved to Cambridge, concluded that the new palaeomagnetic evidence was not indicative of drift. However, his alternative explanation was hardly less unsettling: 'the planet has rolled about, changing the location of its geographic poles'.[53]

Runcorn considered that mountain building or convection currents in the mantle could encourage the polar walkabout. Others were less enthusiastic about reinvoking the horror of convection currents simply to avoid giving sanction to continental drift. One of Runcorn's subsequent papers was rejected with the plea: 'This is another in a number of papers by Runcorn on convection in the mantle and I hope it will be the last'.[54] Gold also saw that his plastic substratum and wandering poles offered an acceptable alternative to drift to explain the curious palaeoclimatic data. His polar shifts had occurred at intervals of about 50 million years. They were the Earth's periodic revolutions, providing sudden phases of climatic transformation and rapid evolution that mark the boundaries of the geological periods.

An identical debate had haunted Palaeoclimatology in Wegener's time. As ever in the contentions over the Earth, the argument was firmly lodged in the psychology of perception. Even the Imperial College palaeomagnetists, who believed that drift was the explanation of their anomalous results, followed the innate and conservative tradition of plotting the pole moving across a map of familiar and fixed geography. Man had always considered the solid rock surfaces as being fixed. To find that an abstraction, a pole, had in the past been located in a different position in relation to this foundation, simply meant that the pole had wandered. Except to the mathematician, rocks will always seem more permanent than the moment of inertia. This contradiction, allied with the term 'polar wandering', dulled the radical nature of the new information.

In 1956 E. Irving at Imperial College had recognised that if the results from Palaeomagnetism and Palaeoclimatology were combined, the drift argument could hope to be fortified.[55] An act of an earth scientist. For the next few years a variety of tests were used to show how the evidence of past deserts, ice-caps, reefs and tropical forests was entirely consistent with the information on latitude coming from magnetics. In 1956 the Cambridge group were themselves beginning to suffer a pole change and soon began to overtake in ambition the London group with their attention only to changes in latitude. Runcorn collected samples from North America in 1956 and by 1959, after encountering Harry Hess at a meeting of the American Association of Petroleum Geologists in Atlantic City, had convinced himself of the opening of the Atlantic.[56]

Runcorn had found a solution to the polar wandering ambiguity. Differences betweeen the polar wandering curves from two separate continents demonstrated the movements of the continents relative to one another. With this neat legerdemain he found that Europe and America shared a common polar wandering curve until the Triassic but since that time the paths had become separated; there had been relative movement. Not only was this the first scientific demonstration of drift on any other grounds than those of resemblance, of coastlines, fauna, geology, but it also provided the spur to a new wave of interest in Wegener's theory.

The new techniques of geomagnetism had spread rapidly to other European centres and to America. Yet the power of fixism in America was not to be broken by a few slick English demonstrations of junctions in the polar wandering curves from different continents. In 1960 when all English geomagnetists believed that their results could only be explained through drift, the influential 'Review of palaeomagnetism' by the Berkeley geomagnetists Allan Cox and Richard Doell[57] still preferred the American orthodoxy. One year later their resistance had begun to crumble: 'It is difficult to explain all of the presently available palaeomagnetic data without invoking continental drift'.[58] In 10 years a new science had grown up that could hope to locate the movement of the crust relative to the reference frame of the Earth's axis. As a new science it could exist independent of the older schools of Geology and Earth Physics with their respective observational and mathematical traditions. It was a union of Geology and Geophysics – a geophysics of rocks at the surface that obtained primary historical and therefore geological data.

The expanding Earth

The palaeomagnetic results surfaced at a small yet important and international conference organised by Samuel Warren Carey at the University of Tasmania in 1956. Carey was born on Wegener's 31st birthday in 1911. His Geology was learnt at the University of Sydney where he also undertook research. A fascination for moving continents arrived while still a student; but it was when he travelled to Papua and New Guinea to undertake an eight year period of work as an oil geologist that his student reconstructions of the globe became reconfirmed in the experience of working with the Dutch geologists in that singular region where the experience of drift was 'self-evident'.

Carey thought that the orogenic belts, the mountain ranges and fast-changing islands of a region like New Guinea were produced by the

Samuel Warren Carey, 1911–
Australian geologist

moving continents and therefore must record their past relative motions. Orogenic belts 'certainly looked as though they had been bent and stretched and disrupted'. Pangaea could be reconstructed with no more evidence than was available through undoing all the collision zones. By 1937 in a draft of his D.Sc. thesis, Carey wrote up this new theory, but had second thoughts about producing it in the final version.

After two years' service as a captain in the paratroopers during the Second World War, Carey assumed a position that provided a geographical ratification of his mental independence. As chief government geologist and soon Professor of Geology at the University of Tasmania, Carey became the lord in his own remote island castle. His first paper from the new isolation is loud in its proclamation 'Tasmania's place in the geological structure of the world'.[59] One thinks of William Lowthian Green, asserting his intellectual independence from Hawaii. Carey thought so too; his identification with Green was unashamed: 'William Lowthian Green' he later wrote 'working alone in Hawaii remote from the conceptual pollution of orthodox establishment'.[60] With his shy birthday partner Wegener there is less empathy. Yet if Wegener mounted a case for drift as an attorney, Carey presented his own ideas as a preacher. The command, the rhetoric, the power, the wit, are those of a great 19th-century natural scientist, presenting his theories from the pulpit, convincing the sceptical through rhetorical hell-fire and damnation.

The Hobart Symposium was as geographically antipodal to the research centres of Europe and America as it was to the orthodoxy that they claimed to represent. The problem in shouting from the bottom of the world is one of being heard. Carey obtained £500 from the Australian Academy of Sciences and one principal guest: Professor Chester Longwell, a veteran of the continental drift debates and scourge of Velikovsky. With the exception of Lester King from Natal, the rest of the speakers and audience were local Australians, or casual visiting geologists. The balance of the symposium can be gauged from the written report[61] in which Warren Carey occupies more than half of the text.

Chester Longwell as chairman was intended to provide respectability and impartiality, but betrayed a sceptical and even sardonic attitude to it all, offering the American contempt of drift through a thin mask of indulgence. Desperate to say something that might belie the depth of his opposition, one can almost hear the teeth tightly clenched as he states 'I am convinced that the arguments on continental drift have been a wholesome stimulus to science'. 'In my view the hypothesis of continental drift has not failed so utterly that it deserves the death sentence.' Antipodean isolation would continue to suffice. Other remarks fulfil Longwell's need to balance his impartiality with mischievous and by now archaic jibes against Wegener's matching of the terminal moraines, and even of Du Toit's 'belief' that as the lines of longitude converged towards the Poles, so continents suffered extension as they journeyed towards the Equator and compression should their motions reverse. Du Toit believed no such thing; instead describing the effect of the ellipsoid on sections of continent; the midriff bulge of the Equator being more tightly curved than the flatter poles. 'I am confident' Longwell concludes his own comments and those of the other contributors 'that some of the published results will find a useful place in the current literature of geology'. That the referee was hostile made no difference to the papers presented. After Irving, now at the Australian National University at Canberra, had talked about the new polar wandering information to have emerged from palaeomagnetism, and Lester King had detailed the reconstruction of the Southern continents 'as plain as a poster' and appealed for some sympathetic Northern assistance, it was left to Warren Carey to propose his own radical theories of continental drift.

If palaeomagnetism offered the first major scientific breakthrough in the continental drift debate, then Warren Carey was offering the most significant conceptual transformation. While all other support-ing evidence for drift had attempted to match disjuncts; broken continents that had once been conjoined, the experience in the Dutch East Indies had inspired Carey to study the deformed belts of the mountain ranges and island arcs in order to read in them the record of the drifting continents. This study had led Carey through a phase of introspection into mounting the first ever critique of the fundamental problem of scale that beset Geology: 'the folds and faults which have so absorbed the attention of geologists since the birth of structural geology are not the first order structures but are of the second and lower orders. We have been so intent on the masonry, the columns and the capitals that we have not seen the temple.'

Carey discusses his proposed system of first order structures, structures for which, by definition, there is no name, a void that he happily fills with a new vocabulary: 'orocline' – an orogenic belt or mountain range with a sudden change in trend; 'rhombochasm' – a parallel-sided gap in continental crust filled with oceanic 'simatic' crust as a result of dilatation; 'sphenochasm' – a triangular gap of oceanic crust where continents have split and rotated around a pivot. The Greek compounds proliferate: 'orotath', 'oroclinotath', 'nematath'. Only 'megashear', a strike-slip fault of great horizontal displacement, has immediate meaning.

The story is pieced together: first the demonstration that 'the Great Rift Valleys of Africa, the Red Sea, and the Atlantic Ocean are similar phenomena, but differ in degree of dilatation.' The extensional rift recently discovered down the centre of the mid-Atlantic ridge is of the same form as the rift valley: 'the dilatation is still proceeding while we debate it'. Before going on to answer the complementary question 'what happens to the ocean floor on the other side of the separating continents?', Carey continues the tectonic arguments, showing how the formation of one of his first order structures, oroclines, spheno-chasms, etc., must require the presence of complementary structure.

118

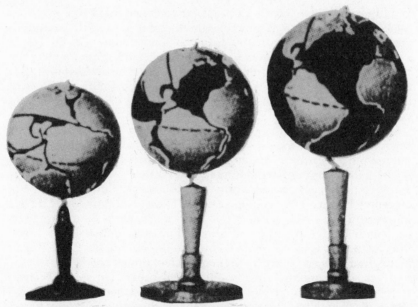

Hilgenberg's 1933 'expanding terellae' – the continents reconstructed on a 3/5 diameter ocean-free globe

He is offering for the first time a suggestion of how movements on the whole globe must be interrelated: 'A megashear must either go right round the globe or begin and end at oroclines or rhombochasms'.

Starting with Alaska, Carey begins to test the pattern of blocks opening and closing. By the time he has reached the match between South America and Africa he has arrived at the original 1920s debate. The final word in that encounter had been left to Jeffreys who had written that the fit of South America into the angle of Africa 'on a moment's examination of the globe is seen to be really a misfit by almost 15°.' Carey had set about exploding this 'shibboleth' of fixism. Through the use of transparent plastic he has moulded the form of a large globe and transferred the outline of South America against the outline of Africa plotted at both the 200 m and 2000 m depth contours; and found that the two match admirably. Reaching for the black ink he scores an underline of emphasis: 'Whether the continental drift hypothesis be true or false, this argument should never be used against it again'.

The Caribbean he explains as a complex region of sphenochasms, rhombochasms, oroclines and megashears; sufficient to allow its total reconstitution within any closed Atlantic reconstruction. On pondering the history of the Mediterranean he begins to explain the greater pattern: a shear system with left-handed motion passing east–west between Europe and Africa is but one section of a global 'Tethyan Shear System' that rings the world, passing through the Caribbean and the East Indies. Such a global shear zone had first been proposed by William Lowthian Green for whom Carey had such great geographical and mental empathy. Carey is almost home. For the Dutch East Indies is where his own global thinking had begun, and like the Dutch he believed the region to be at the focus of the understanding of the Earth, even offering some antipodean answers to the problems of European geology: 'In the East Indies of today we may see the Alps of yesterday.'

'As a geologist my duty is therefore to follow where the orocline interpretation leads me': so Carey ends his paper. Yet surely there has been some deflection in Carey's path, his thought is oroclinal, for his initial statements denied the fundamental scale of methods of analysis employed by geologists and now he wishes to defend his course of argument based on the 'geologist's duty'. What has happened? He has discovered that much of the morphology of the Earth's surface can be explained by the opening of rifts and chasms. Most of his novel interpretations were to become confirmed a decade later: the Bay of Biscay had opened where Spain had rotated around the Pyrenees; the Red Sea was a mature rift, an adolescent ocean. However, Warren Carey, in his inditement of the scale of geological theorising and in his creation of a new scale that was truly that of the whole spherical Earth, had on the way made some observations that he believed to be value-free, neutral of assumptions. These observations concerned the major morphological features of the Earth for which he had provided the new vocabulary. All of those that involved ocean crust were extensional. Therefore, in compiling these features from across the globe, Carey had constructed out of extensional building blocks an extensional globe. Like the pre-war German theorists Carey has seen the expanding Earth that lies implicit within Wegener.

In 1953 Carey had sent a paper on convection cells and continental drift to the American Geophysical Union, and after this was turned down, his radicalism fully confirmed, he became an expansionist. The problem first arises with a contemplation of the Pacific Ocean. For the Atlantic to open requires that the Pacific should be closing, yet Carey found his rhombochasms and spenochasms scattered all around the periphery of that mighty ocean: 'far from shrinking during this time, the Pacific also has expanded greatly'. There has been universal dispersion; all blocks have moved apart relative to one another, at a rate accelerating through time. As for a cause he is prepared to admit that it lies outside his field of competency.

Warren Carey holds a unique position within this history because he was the first geologist to challenge the most fundamental understanding of the scope and scale of investigation, and to erect in its place a new set of larger-scale structures organised in a global configuration. He achieved this realisation from working outside, in semi-isolation from the northern mainstream. As a Southern Hemisphere geologist, continental drift was itself self-evident; his own critique of Geology was the product of a unique vision and analysis. This conception of the expanding Earth was not to be perturbed by any of the discoveries of the next two decades and his articulate enthusiasm for the idea has been sustained through to the present. His formidable powers as a orator won him many converts from the first presentation of his *'radical* even *shocking'* views, as Chester Longwell termed them. More than a quarter of a century later, it is common to find a paragraph or chapter in an Australian geological publication extolling the expanding globe. Carey's concluding sentence at the Hobart Symposium is revealing of the continuing self-validation of Geology relative to the other sciences: 'the orocline induction could be the clue leading to new fundamental developments in physics.'

6

A naval engagement

The birth of Oceanography

Some of the key skirmishes in the crusade for continental drift had their specific geographical locations: the Dutch East Indies with their 'self-evidency' of huge lateral forces and colliding continents; the Alps built out of great sheets of sediment that had slid over and become piled one atop the other; the East Africa rift valley that gave hints of tension when all processes were thought to be compressional. Yet the final and conclusive battles concerned problems that were out of range of the land and therefore of geologists: naval engagements fought by American and British vessels in the middle of the world's great oceans.

The heritage of navigation that these two nations shared had sponsored oceanographic exploration through the 19th century. In 1854 the American Matthew Fontaine Maury published the result of his voyages: the first crude contoured map of a whole ocean basin (the North Atlantic), with a mysterious central shoaling: the Dolphin Rise. Charles Darwin had employed his time on the *Beagle* musing on all manner of geological and zoological problems, and wrote in 1842 a book *The structure and distribution of coral reefs* which claimed the Pacific atolls were coral towers, slowly erected to keep level with the sunlit water, around the sinking flanks of volcanoes that were now lost hundreds of metres beneath the surface.

The first deep-ocean expedition left England aboard the *Challenger* in 1872. Among its findings was that Maury's Dolphin Rise was a long and sinuous mid-Atlantic ridge, a discovery that within a few years had become associated, even in the pages of *Nature*, with the lost Atlantis. Such ridges were also to provide the foundations for the land bridges. After *Challenger* the oceans passed out of the thrust of scientific enquiry. The vast and inaccessible ocean floor could not be understood through scattered samples clumsily and randomly collected in dredges, from a ship moored a few kilometres overhead. Mapping the ocean floor had to await the creation of the new technologies of remote sensing.

Echo-sounding was first developed to detect icebergs in the aftermath of the sinking of the *Titantic* in 1912. After 1914 there came to the North Atlantic a more menacing threat than icebergs; the German U-boat provided the motivation for the first period of military investment in scientific research. By 1915 10 Canadian-built British submarines had been fitted with echo-detectors. The US Navy was encouraged to fund two research centres and, though the bonds made between scientists and the admirals lapsed at the end of the war, the precedent had been forged and the co-operation was to be resumed and massively fortified only two decades later. As the research teams were disbanded, so the technology trickled out to be used for scientific oceanography. In 1922 the USS *Stewart*, equipped with the new deep-water sounding device, sailed for Gibraltar, taking, in a nine day trip, 900 depth readings. This was a considerable breakthrough: the *Challenger* in 3½ years had achieved only 300 deep-water, piano-wire soundings.

The first important scientific use of the equipment came with the German Atlantic Expedition of 1925–7; the ship *Meteor* carried two depth finders that produced a total of 33000 duplicate soundings. In 1924 the US Navy had attempted to launch a large oceanographic expedition but Congress had no intention of funding what it could only perceive as unnecessary naval jaunts in the spurious name of science. In a number of European countries the naval support for Oceanography was more forthcoming. While the echo-sounding programme had been initiated primarily to enable the detection of submarines, it was the possibility of utilising submarines as research platforms that attracted a Dutch geophysicist, Felix Andries Vening-Meinesz, to begin measuring gravity at sea. Vening-Meinesz studied first as an engineer, but for much of his life was Professor of Geodesy and Geophysics at Utrecht and Delft. Early in his career he designed a multiple pendulum gravimeter; the first that could cope with even slight movements of the observation point, and in 1923 took it on board one of Her Majesty's submarines where at depths of 80 m, in the quiet zone of the oceans, he gained the first offshore gravity readings.

In 1926 Vening-Meinesz and his instruments hitched a submarine-ride from Holland to its Eastern Colonies via the Panama Canal. These observations confirmed that the condition of isostasy applied equally over the oceans as over the continents.[1] This was the most important scientific confirmation of the impossibility of land bridges. It was generally ignored. More significantly Vening-Meinesz found that all the deep-ocean trenches that he traversed in his voyage to Java were also the site of large gravity deficiencies; isostasy had not been attained over these linear deeps.

By 1932 Vening-Meinesz was prepared to extend his thoughts on the existence of these deep-ocean troughs.[2] The network of seismic instruments in operation through this region had revealed that seismic activity on the landward side of the deeps continued down to several hundred kilometres; far deeper than the few tens of kilometres

Felix Andries Vening Meinesz,
1887–1966
Dutch geophysicist

that was the lower limit on earthquakes over continental regions. The gravity deficiency could only be sustained if the ocean floor was somehow being dragged down. The downbuckling Vening-Meinesz associated with convection currents in the substratum. Such currents, he thought, were brought into existence through the thickening of continental crust during episodes of folding. The continental rocks had a higher level of radioactive heat generation than their oceanic counterparts and this new heat source in time caused the substratum to engage in temporary 'convective overturn' which would complete about half a revolution before re-establishing a new equilibrium. Sporadic convection could account for former orogenies 'and was not necessarily in contradiction with the main subject of Wegener's theory'.

In 1927 the US Academy of Science appointed a committee on Oceanography to investigate the rumours that American research lagged behind that of the Europeans. As a result of the committee's findings, Federal funds were allocated for a variety of projects and in 1930 with 3 million dollars of the Rockefeller fortune, an oceanographic research institute was founded at Woods Hole, Massachusetts. At the same time, an Australian explorer, Sir George Hubert, had gained from the US Navy the use of an elderly submarine for a planned visit to the North Pole. The success of Vening-Meinesz's submarine geophysics led the Americans at Woods Hole to equip the ship with gravimeters and magnetometers, but after a series of breakdowns the vessel got no further than the edge of the continuous ice.

To avoid further disappointment, the Dutchman himself was invited to join a combined Navy-Princeton Gravity Expedition of 1932, attracted by the opportunity to investigate a deep-ocean trench in the Caribbean. Among the Princeton scientists on board there was a graduate student, Harry Hammond Hess, who was instructed to learn both the theory and practice of the science of submarine gravity in order to interpret the expedition's results once Vening-Meinesz had left. Thus Hess became Vening-Meinesz's pupil, not only in the craft of gravity but also in the theory and philosophy of a convecting substratum. In the absence of a research submarine, Hess took the unprecedented step of joining the Naval Reserve in order to continue his marine investigation of the Lesser Antilles.

The thousands of echo-soundings across the mid-Atlantic ridge obtained in the mid–1920s, by the German vessel *Meteor*, provided the first accurate picture of the topography of a submarine mountain range. In 1933–4, a British expedition set out to study another such chain of mountains: the mid-Indian Ocean, or Carlsberg, Ridge. As the ship *John Murray* traversed across the spine of the ridge, the echo-sounder showed the ridge crest to contain a deep gully. This central trough within the sinuous ridge was compared by the expedition leaders (J. D. H. Wiseman and R. B. S. Sewell) with the topography of the domed sides and central gully of the great rift valley of east Africa.[3] They had also noticed something strange about the seismicity of the region. The earthquake centres ran 'down the middle of the Arabian Sea where we know that the Carlsberg Ridge runs its course. A similar belt can be traced down the length of the Mid-Atlantic Ridge'. The oceans were beginning to reveal far greater order than the continents.

Field's 'children'

The great automobile boom of the 1920s had inspired an oil rush sending prospectors wildcatting over the dusty lands of the American South. The successful oil company was the one that could increase the odds of finding oil in any expensive drilling venture. To obtain knowledge of the buried geology without the expense of a drilling rig was every oilman's dream. Seismic prospecting emerged out of the application of the new seismology to problems of sound-ranging the big-guns on the Western Front in the First World War. On the British side were physicists (many from Cambridge University) led by Captain Lawrence Bragg. For the Germans there were seismologists, including the Gottingen Geophysical Institute's founder Emil Wiechert and one of his students, Dr Ludger Mintrop, who only convinced their general command to use mechanical seismographs to locate enemy guns in 1917. The same year the US Bureau of Standards brought together a team of sound-ranging physicists including J. C. Karcher. The war over, these three teams began to compete with one another over the application of these seismological techniques for mapping subterranean structure. The first experiments with commercial geophysics were undertaken in America. The return signals from a surface explosion were recorded at a number of locations and the equipment was then shifted until a crude picture of the differing seismic velocities of the buried strata could be constructed. The first successful use of seismic prospecting came in 1923 when the German physicist Dr Ludger Mintrop adapted some military equipment and, through the mediation of the Dutch oil geologist van Waterschoot van der Gracht, brought his skills to be used by the Marland Oil Company operating in Oklahoma and coastal Texas. Having effected the mediation of one aspect of German geophysics for the Americans, the Dutchman was to try again with the promotion of Wegener's drift in the American Association of Petroleum Geologists symposium of 1926.

Around 1930 a tall, energetic, Texan physics graduate student, Maurice Ewing, obtained a summer job with an oil company to assist in the search for oil under the swampy lakes and creeks of the Louisiana coast. Ewing learnt how to fix home-made bombs into the mud, detonate them and record the return signals on the simplest portable seismograph.

The summer geophysics job steered Ewing into his career; the previous years he had been working in the grain elevators. Born in 1906, the oldest child of poor homesteaders, gaining a meagre living in the plains of north Texas, he had been fortunate to escape from the hardship first to the Rice Institute in Houston where, under the influence of an ex-Cavendish physicist and a couple of Harvard mathematicians, he changed from Electrical Engineering to the uncertain science of Physics. Ewing completed a Ph.D. at Rice, and in 1931 presented a paper on the physics of seismic prospecting for petroleum at the tiny annual meeting of the American Geophysical Union. At that meeting was a geologist of unprecedented vision: Professor Richard Field from Princeton.

While in the drift discussions of the 1920s great attention had been given by geologists to defining the status of the oceans, none of them had thought to plan or even consider how such status could be verified

Richard Field, 1885–1961
American geologist

from scientific exploration. Geology was by definition the study of the land-surface of the continents, the oceans occupying three-quarters of the globe were geologically invisible. Field wished to redefine the science. Born in 1885, he had been a student at Harvard under the inspired teaching of Alexander Agassiz, the son of the great Swiss geologist and ichthyologist, Louis Agassiz, who had been the first to recognise the evidence for a former European ice-cap and subsequently the significance of ice ages. While the father was a Harvard professor, the son Alexander had roamed the seas and oceans, extending knowledge of marine life. On 5 May 1881 Charles Darwin wrote to Alexander in order to attempt to encourage some marine geology: 'I wish that some doubly-rich millionaire would take it into his head to have borings made in some of the Pacific and Indian (ocean) atolls, and bring home cores for slicing from a depth of 500 or 600 feet [150 m]'.[4] Darwin wished to settle the tiresome argument that had been continuing for more than 40 years over the origin of coral atolls. Alexander Agassiz's marine enthusiasm, focused perhaps by Darwin's requests for atoll drilling, became, claimed Field, his inspiration for demanding that geologists should explore the ocean floors.

Field's vision did not communicate itself through scientific endeavour but through teaching. He taught Geology, first at MIT, then at Brown University, and with such success that he was invited to Princeton to present the general Geology courses. His enthusiasm was too great to be restricted by the 19th-century frontiers to the science. Soon after arriving at Princeton he organised the first of a series of annual summer geological '*Field*' trips, hiring a Pullman dining car to roam for 10000 miles through the USA and Canada, taking both students and, as the fame of the venture grew, visiting geological eminences. During the late 1920s he became a close friend of Major William Bowie, the great defender of Dutton's isostasy at the 1926 continental drift symposium. Isostasy depended on gravity observations for its confirmation; it was a suggestion from Bowie that led Field to invite Vening-Meinesz to Princeton and to obtain from Francis Adams, Secretary to the Navy, the use of the submarine. The encounter with Vening-Meinesz provided Field's first insight into the means by which the oceans might be studied. He became an active member of the newly formed, 200-member American Geophysical Union and headed its committee for exploration of the ocean basins. Yet his research was not into ocean geophysics but into locating students of promise who might be infected with his enthusiasm and might go out to carry on his mission. At this research he was the undisputed master. His parents had been prominent in the theatre; his wife was the daughter of a famous actress. Whether as a lecturer or as a politician in obtaining funds, his rhetoric was as of an actor, his authority that of the director in affecting the course of his students' lives.

In listening to Ewing's 1931 talk he saw that the new seismic prospecting would allow it to be the mediator in the extension of

Maurice Ewing, 1906–1974
American geophysicist

geology into the oceans. In 1934 he visited William Bowie, Director of the United States Coast and Geodetic Survey, and obtained from him the promise of a $2000 grant to finance some preliminary offshore geophysics. On a snowy November day they travelled together to Lehigh University in Bethlehem, Pennsylvania, where Ewing had a job as a desperately overworked instructor in Physics. Ewing had been undertaking geophysical research on whatever problems and with whatever resources were available in the Lehigh Valley. The true nature of Depression science is to be found in publication titles such as 'Locating a buried power shovel by magnetic measurements'. To save money on explosives, seismic measurements were made from the valley's own quarry blasts. Bowie and Field asked Ewing whether he thought the seismic prospecting techniques could be used to find the nature of the continental shelf. Ewing was immediately enthusiastic, demanding only equipment and a ship.

Field had originally wanted to bring Ewing to Princeton but failed to convince the university to provide support. In 1934 Harry Hess who had worked with Vening-Meinesz on submarine gravity obtained a Princeton lectureship in Mineralogy, for although he had become involved in Geophysics under Field's influence he was foremost a petrologist prospecting between 1927 and 1929 in Rhodesia (now Zimbabwe), and researching a peridotite from Virginia for his doctorate. Field hoped to extend his students' knowledge of Geophysics through introducing them to Ewing, and on several weekends they would all drive the 150 km to Bethlehem to assist with the onland section of the seismic traverse that was intended to run across the New Jersey coastal plain before moving out to sea.

A mixture of friendship and rivalry affected the relationships between Ewing and Hess. Ewing was exactly 12 days older than the Princeton petrologist; both had started as students of Electrical

Engineering. In time, the jealousies and bitterness were to become overwhelming.

In 1935 maritime seismic reflection was ready to be launched. Ewing set out to extend his land-based line of seismic recordings offshore from the Virginia coast on the Coast and Geodetic Survey's vessels *Oceanographer*. Yet a change in captaincy forced him to co-ordinate his experiments with a strict hydrographic schedule and before he could fully adapt the land-based techniques to the new range of problems of working at sea, the cruise was over. Enormously frustrated, but with the certainty that offshore seismic reflection profiling was about to be possible, he went to visit Field for counsel as to his next move. Field sent Ewing to Woods Hole where the Texan bomber could hope to convince the directors to offer him cruise time on one of their research vessels. First the oceanographers required some evidence that the arsenal of explosives would not endanger the lives of the scientists or crew. This surety given, Ewing set off with a whaleboat full of small packages of gelatin that were fired on the sea floor as the wherry pulled away from the research vessel *Atlantis*.

The results were extremely promising; the thick piles of sediments that lay beneath the continental shelf were beginning to reveal themselves. Each summer Ewing returned to Woods Hole until by 1940 he had improved the most difficult part of the enterprise, the bomb-letting. Out went the whaleboat and in came timing and flotation devices which could be recovered after the sea-floor explosion had been triggered. Remarkably, there were no serious accidents: the TNT was stored in the chief scientist's cabin, packed as a powder into meteorological balloons, and drilled to be fitted with blasting caps on deck in all weathers. Alongside the deck leaky flotation tanks were lashed to the rail, each filled with 30 gallons of high-octane gasoline.

In these Depression days, Field insinuated his new world-view into the lives of four young men who were to effect transformations in

Harry Hammond Hess, 1906–1969
American petrologist and
geophysicist

the science of Geology, far more radical than Field himself could hope to see. The third of these transformers of science was an English geophysicist, Edward Bullard, whom Field first encountered in 1936. Bullard later wrote 'Field was in a large degree the founder of marine geology. He explained to me that what was wrong with geology was that it studied only the dry land and that you could not expect to have sensible views about the earth if you studied only one third of its surface.'[5] Field saw exactly how Geology should extend its dominion over all the face of the Earth: 'the critical problem was to study the ocean floor starting from the land and working outwards into the ocean.' Bullard, like Ewing, was inspired by Field's enthusiasm for such an endeavour: 'the burning zeal of an Old Testament prophet. He could not take no for an answer, he would not stop talking, he had no doubts, he was embarrassing and sometimes a nuisance, and yet he struck the match that set earth science alight'.

Edward Crisp Bullard was to become the co-founder with Maurice Ewing of the modern science of Marine Geophysics. Born in Norwich in 1907, he first took an interest in running his family's brewery before going to Cambridge where he obtained a first in Physics, and then continued to undertake research in Nuclear Physics at the Cavendish Laboratories. In 1921 the Director of the Survey of India, Sir Gerald Lenox-Conyngham, retired back to England and convinced Trinity College in Cambridge to offer him a fellowship to pursue his desire to train a new breed of surveyors. The university agreed to set up a Readership in Geodesy, a position to be held by Lenox-Conyngham, but without salary, and in 1931 a second Readership was established in Geophysics, this time stipended, and offered to Harold Jeffreys. A new Department of Geodesy and Geophysics was created and Lenox-Conyngham, realising that he required a counterweight to Jeffreys, someone cognisant in the new experimental Geophysics, one Sunday at Trinity high table, positioned himself next to Sir Ernest Rutherford, and asked the great physicist if he could recommend a candidate. Rutherford proposed Bullard and then warned Bullard of the difficulties of obtaining any appointment in these times, sufficient to convince him to change the course of his research and take up the appointment.

Bullard set off to perform the fieldwork for his new Ph.D. on the gravity field over the East African rift valley. After several years learning the techniques of his new science, 'of applying physics to the earth and trying to understand the modes of thought of geology', Bullard began to realise 'how weak the underpinning of geological hypotheses was'.[5] His first research took geophysical techniques to the most interesting and complex parts of the Earth's surface – the gravity mapping in East Africa led him into another survey around the extraordinary spread of dense peridotite rocks across Mount Troodos in Cyprus. At the same period he became interested in measuring the variations in continental heat flow from one region to another. Instead of utilising geophysical techniques simply to solve geological prob-

John Tuzo Wilson, 1908–
Canadian geophysicist and Earth
scientist

lems, he wished to redefine the whole form of the investigation of earth problems and history. It was in this revolutionary spirit that he encountered Richard Field, who invited Bullard to America in 1937 '(and I suspect paid my fare), he drove me hither and thither in his car, tried to teach me geology, and sent me out to sea with the Coast and Geodetic Survey and with Maurice Ewing. After a few days he was preaching to the converted but he did produce a sense of urgency that was new to me in science.'

The fourth of Field's protégés, John Tuzo Wilson, was to have no connection with marine work for many years to come. Wilson was born in Ottawa in 1908, and at the age of 15 was sent to the northern woods for the first of five summers working in a forestry party where he later became a field assistant to the Cambridge geologist, Noel Odell, recently returned from the famous attempt to climb Mount Everest from Tibet. Having already completed a first year in Physics, Wilson was inspired by Odell to choose an outdoor life in contrast to the 'repetitious and stuffy' practice of Physics.[6] His professor was 'dismayed and irritated that any promising student should abandon so prestigious a subject as physics for geology, then held in very low regard.' Neither did the geologists want a physicist in their descriptive midst. A compromise was arranged by the most enterprising of the geologists, Lauchlan Gilchrist, who was one of an investigative committee set up by the US Bureau of Mines and the Geological Survey of Canada to study and test the new methods of geophysical prospecting.

By 1927 Gilchrist had been sufficiently impressed by the possibilities of Geophysics to initiate a course in mining prospecting, and encouraged Wilson to take a double major in Physics and Geology. However, there was still hardly any expertise in Geophysics; two very incomplete lecture courses were organised, at one of which he was the

131

lone student. At the end of his degree Wilson could claim to be as close as was then possible to being a Geophysics graduate.

Wilson won a scholarship to Cambridge University, where he attended the lectures of the master of Earth Physics, Harold Jeffreys: 'Unfortunately I could neither hear nor understand them, but Jeffreys was accustomed to this and did not hold it against me.' The two years spent in Cambridge were filled with a certain idleness; there was no simple explanation of what he was supposed to be doing there, learning about a new science that there was no one qualified to teach; Edward Bullard had slipped away on his new appointment without encountering Wilson.

On return to Canada, there was no regular work available. The Director of the Geological Survey suggested that Wilson bridge the Depression; advice that encouraged him to go to Princeton, to undertake a doctorate under Field. The chosen research area was in the Beartooth Mountains of Montana. Wilson was introduced to great faults, geology from aerial photography and a delight in the wilderness. He was also overtaken by Field's vision: 'He bubbled with energy, enthusiasms and powers of persuasion; he had a sweep of imagination lacking in his research-conscious colleagues.' On return from Princeton, Wilson was offered the promised job in the Canadian Geological Survey and was to spend the next 10 years involved in mapping programmes interrupted by army service. As it turned out, he was biding his time.

Research on marine geophysics in Britain was slow to start: only in 1936 were preparations made for taking one of Vening-Meinesz's pendulum gravimeters in a submarine out along the continental margin to the west of the British Isles. However, by 1938, Bullard, with a small Cambridge team, had begun to emulate Ewing's elaborate and difficult procedures over the British continental shelf. By the following summer, in collaboration with T. F. Gaskell, he had improved on Ewing, placing both explosives and geophones close to the sea surface rather than resting on the sea bottom. This is the method of offshore seismic prospecting that, with many technical improvements, has been in use ever since. Both Bullard and Ewing now wished to take their techniques out into the deep ocean.

Deep-sea Geology had yet to be born: in the 1946 encyclopaedic *The oceans* it was remarked that 'From the oceanographic point of view the chief interest in the topography of the sea floor is that it forms the lower and lateral boundaries of the water.' Before they could attempt such missions, the deep oceans had again become out of bounds; the submarines were back, only now with a cargo more deadly than gravimeters; research shifted from pure science to the military.

The war boom

Richard Field became a victim of the war even before it had started. In

1938 he was elected President of the American Geophysical Union and was set to be the host and leading scientist when the International Union of Geodesy and Geophysics held its Seventh General Assembly in Washington, DC. As the representatives gathered, Germany invaded Poland. Most of the foreign delegates had already left by the opening session. Five years later he suffered serious head injuries in a car crash from which he never fully recovered. He died in 1961, his obituary for the American Geophysical Union written by Harry Hess.[7]

The transformation from the poverty of the 1930s to the massive research funding of the war years was rapid and profound. In the spring of 1940, while America's involvement was still uncertain, the Government set up the National Defense Research Committee to provide the military with immediate access to scientific expertise through contract research programmes. By the time of Pearl Harbour, the National Defense Research Committee had 450 contracts involving more than 2000 scientists; including all oceanographers (bar a few marine biologists). The total money spent on war research rose from $100 million in 1940 to $1.6 billion in 1946, and in 1942 the Woods Hole Oceanographic Institution had to withhold the annual treasurer's report in case it should hint at 'the magnitude, if not the nature of the investigations being undertaken at the request of the government.'[8]

Those private and university research institutes that had weathered the lean years now provided the nucleus of enormous establishments; Woods Hole grew from a summertime staff of 60 to almost 350 full-time employees. On the West Coast, the Scripps Institute of Ocean-ography (founded privately around the turn of the century and incorporated into the University of California in 1912) also enjoyed a massive expansion. These new employees had no training in the demands of oceanographic research and thus did not immediately serve to increase the scientific productivity. Instead it was those who had pioneered the science before the war who suddenly found themselves lording over great research empires. The ambitious deep-ocean geophysicists Bullard, Ewing and Hess had very different but highly successful and formative war careers.

Within a few days of the outbreak of the war the German navy and airforce began to lay magnetic mines on the sea bed around the ports and harbours of the east coast of England. The magnetic mine was a British invention, first produced in 1917. The success of this reintro-duced technology was dramatic. Within two months of the declaration of war 29 merchant ships and 1 destroyer had been sunk by magnetic mines. In November 1939 over 200000 tonnes of British shipping was lost. It was not until 21 November that the first intact specimen of the unknown weapon was recovered. A team from the Admiralty Research Laboratory, which by now included Edward Bullard, was placed in charge of countermeasures. By the summer of 1940 the success of this operation was so complete that the Germans were led to redeploy their sophisticated weapons as bombs in the London blitz. Bullard had gained his first experience of marine magnetics, and also of the

Sir Edward Crisp Bullard, 1907–1980
British geophysicist

possibilities of rapid research and development. His reputation was established and in 1944 he was promoted to be Assistant Director of Naval Operational Research under the leadership of the physicist Patrick (later Lord) Blackett. In the final stages of the war he was serving on the Committee studying the threat from the German V1 and V2 bombs.

Bullard spent much time in this final war period discussing with Blackett the potentialities of a post-war science. The initial experience of magnetic mines had had a profound impact on Bullard, extending his geophysical interests into a whole new area. One of the problems they identified as requiring much further research was the origin of the Earth's magnetic field. At the war's end Bullard returned to the unhurried tranquillity of Cambridge University as a Reader in Experimental Geophysics and Head of the Geophysics Department, but found the calm inertia and the post-war research poverty so stultifying, that in 1948 he left England to become a Professor of Physics at the University of Toronto, where he encountered Wilson.

Maurice Ewing was called back to work at Woods Hole on war-related research soon after he had completed his 1940 summer cruise. Initially without salary, he worked with the Woods Hole director Columbus Iselin on problems of sound transmission in the oceans. Within a few months the US government, sensing an impending American involvement in the war, offered one of the first wartime marine research contracts. Like Bullard, the new experience of wartime problem-solving turned Ewing into a leader, and he thrived on the responsibilities of organising and liaising with the Navy. His new reputation led Columbia University in 1943 to offer him a chair in Geophysics; a position that he accepted on the termination of his war research in 1944. The uptown Manhattan building that was to be his Department and into which he was ushered on arrival had 'no floor and the worst collection of broken furniture you have ever seen'.[9] His three small rooms in the basement of the Geology Department had formerly been used in the development of the atomic pile that had given the wartime Manhattan Project its uptown origins.

For four years Ewing was encumbered with a heavy teaching load and a rundown building. He was on the point of quitting when the university was offered, and accepted from a wealthy banker, Thomas Lamont, a handsome estate, situated high up on the Hudson Palisades, which was presented to Ewing to be turned into a Geophysics centre. He arrived with half-a-dozen graduate students in 1949 and set about turning the greenhouse into an instrument shop, the seven-car garage into a storeroom for deep-sea cores and the covered swimming pool into a test tank. Down in the cellar he installed a group of seismic recorders. The Lamonts' bedroom became Ewing's office and he later captured their breakfast room. The rose garden, however, remained and remains a rose garden.

After an initial grant of $500 000 the Lamont Geological Observatory was to be self-supporting from research contracts, grants and endowments. This was made possible because the wartime research funding did not return to the pre-war neglect. As measured by the military funding of science, the Second World War never ended. Ewing's special interests were in the transmission and recording of seismic waves, both from natural earthquakes and from his supervised marine explosions. Through a consistently high level of funding which Ewing was most successful in gaining for Lamont, his institute became a world leader in the geophysical exploration of the deep sea. Expanding at the rate of 20 new staff each year, soon 'drawing boards were set up over bathtubs and file cabinets over toilets'. As space became limited so it became expedient for accommodation and good for research to keep a sizeable contingent always at sea.

Harry Hess had a more direct involvement in the war effort. The morning after Pearl Harbour he left the Princeton Geology Department and took the 7.42 a.m. train to report for active duty where he was first assigned to New York in order to organise detection of U-boats in the North Atlantic. Within two years the East Coast submarine threat had been effectively beaten and Hess became Commander of the attack transport USS *Cape Johnson* in the Pacific. Apart from the occasional real or imagined threat from Japanese submarines and planes, Hess was left free to pursue research in parallel with his duties. Although he participated in the landings on the Marianas, Leyte, Linguayan Gulf and Iwo Jima, this was not a front-line vessel but a support ship; Hess had insisted on the vessel being equipped with the most powerful echo-sounder available and his research consisted of running this instrument almost continuously throughout the ship's journeys across and around the Pacific.

He plotted his courses so as to survey new areas, or else return to some feature that required further scrutiny. Where time permitted, he could complete a partial survey of some curious topography; a lone scientist in the midst of a war, steering his ship to traverse invisible mountains rising a thousand metres or more up from the ocean floor. By the time of the Japanese capitulation Hess had mapped about 100 flat-topped submarine mountains; shaped, he thought, like volcanoes with their tops truncated.[10] They were sprinkled in a broad band running south-east several thousand kilometres from Guam and the Mariana Islands, following a common course to the atolls of the Darwin rise. At first Hess followed Charles Darwin's own thinking in claiming that these sea-mounts were once coral atolls that had failed to keep abreast with sea level as their supporting volcanoes sunk, but then he changed his own arguments to be more in keeping with the lesser dynamism of Geology; they were Precambrian, ancient, more than 500 million years old. He named them 'guyots' in honour of a Swiss American, the first Professor of Geology at Princeton.

Through his direct participation in the war, Hess, unlike Bullard and Ewing and other leaders of the oceanographic institutes, had failed to build an empire and was therefore left to investigate the vast expanse of the oceans alone. At the war's end Hess remained in the Naval Reserve and employed the annual opportunity afforded by his naval visits to study all the echo-sounding profiles sent back to the Naval Hydrographic Survey, in an attempt to find more guyots. Mysteriously they seemed to be confined to his Pacific war supply lines. Oceanographers at Scripps in California had voyaged out to collect grab samples from the table-top summits of the sea-mounts. They could find no sediments older than the Cretaceous geological period.[11] There was something very peculiar beginning to be discovered about the Geology of the oceans; with very few exceptions it was all proving to be extraordinarily young.

Mapping the oceans

While all the preparations for the study of the geophysics of the oceans had been made during the 1930s, it was not until the 1950s that the harvest of the deep began to be brought home. Lamont, under Ewing, was the most vigorous of the three big US oceanographic institutes in the effort to capture large quantities of geophysical data from the deep sea. In 1953 Ewing prevailed upon Columbia to buy a three-masted auxiliary schooner, the *Vema*, originally built as a pleasure yacht in a Danish shipyard in the 1920s, with which to complement his Lamont estate. The toys of the aristocracy became the tools of science. Like the swimming pool and the seven-car garage, the *Vema* was converted for geophysical research, and with a steady supply of funds, and naval surplus depth charges, by 1956 several hundred deep-sea seismic refraction lines had been shot.

The explosion shock waves were large enough to penetrate to the base of the ocean crust, to reflect and refract off the underlying mantle. They revealed a structure of incredible regularity; prior to the war it had been assumed that the Atlantic crust was somehow intermediate between that of the 20 mile (30 km) thick continents and that of the Pacific Ocean. In 1949 Ewing, in collaboration with a research student Frank Press,[12] had shown from the evidence of the surface waves that travelled across the oceans from distant earthquakes, that the Pacific and the Atlantic oceans were identical; their crusts extremely thin. By 1956 Ewing and his researchers[13] had confirmed from the evidence of seismic refraction experiments at sea that ocean crust was about four miles (6 km) thick, across all the oceans. The 19th-century belief that the Atlantic was somehow contrasted to the Pacific was finally exploded.

There remained the mid-Atlantic ridge; was it perhaps a range of fold mountains like the Alps, a series of great fault-scarps like the Basin and Range of Nevada or a great cluster of volcanoes heaped one atop another? Ewing's expeditions had provided samples cored from the flat abyssal floors and on one of his earliest post-war cruises back in 1947, dredged from off the tops of the jagged mid-ocean arêtes samples of fresh basaltic rocks and serpentinite. In 1953 the Royal Society of London held a discussion on the subject of the floor of the Atlantic Ocean to consider the new results.[14] The geologists were also present; G. M. Lees asserted that solid 'geological evidence' must be contrasted with the evanescent 'geophysical conceptions'; and that foundered continents would one day be found interred beneath the lavas and sediments of the ocean bed. Bullard replied, patiently pointing out the new evidence supporting the continuity of the continental Moho with the boundary at the base of the thin crust of the oceans.

Hess also produced a paper at the meeting, offering several hypotheses for the origin of the mid-Atlantic ridge. Freed from the bureaucracy of a large research organisation, and from the possibilities of himself amassing new ocean data, Hess was left to be the most imaginative of Field's protégés. Somehow his interests as a research student had become combined: among the 1954 hypotheses he suggests that a rising convection current is located in the mantle; a mass of basalt swells up and breaks through to the surface carrying huge blocks of mantle peridotite within it. The ridge is raised for three reasons, the upward convection, the expanded high temperature basalt and through the process of 'serpentinisation'.

Hess the petrologist had one advantage in contemplating the behaviour of the ocean floor, in that he was well versed in the chemistry, properties and interactions of the rocks themselves. The theme of his doctoral dissertation was to recur in his ideas and to become for him the most significant transformation in the behaviour of the outer skin of the Earth. Peridotite (from the French for the mineral olivine, *peridot*), a dense rock made principally from that olive-green magnesium silicate, became converted through the addition of water at 500°C into great masses of less dense serpentinite. By 1955 Hess believed that the mantle was made of peridotite and that this had become hydrated to form the ocean crust of serpentinite.[15] Still, in 1954 he employed convection no more liberally than did Vening-Meinesz, only shifting it to a different part of the ocean. There is no indication that Hess had read the earlier, more daring ideas of Holmes.

In the early 1950s Ewing kept his ships at sea throughout the year, collecting all possible data, from sediment cores to geomagnetics, working, and forcing all others around him to work every day, all those hours in a day when exhaustion from sleeplessness did not overcome him. Lamont expanded to be the most important centre for information on the deep-sea floor. In time this would mean that any new theory would have to be tested and approved there. Lamont had become the patent office for inspiration. While a leader in equipment innovation, and with 'an intuitive sense of what is important in science' this bullish desire to amass information and the success with which he gained funding in order to expand Lamont's empire enforced for Ewing the same sense of authority that had possessed Chamberlin. The drama that was to unfold over the next 15 years was to be full of anger and disappointment, for Ewing, the creator of the world's most productive Marine Geophysics Institute was not only a demagogue but also a fixist.

Under the influence of Walter Bucher, Columbia University had become the American flagship for the creation of a greater Geology that was to be more scientific, less ruffled by the breezes of some foppish theory as continental drift. Bucher was Ewing's mentor in whole-Earth theories; Ewing in turn was to impose his will upon Lamont. He had strong views on geologists:

> Annoying fellows . . . spend their time poking around trying to explain this or that little detail. I keep wanting to say, "Why don't you try to see what's making it all happen?" It's as if you set a battleship in front of these men in dugout canoes and after a few years they found out how you get the lights to go on.[9]

While he could patronise geologists as ignorant savages, the advocates of continental drift required a sterner rhetoric. 'Ewing' one of his contemporaries recalls 'was the leader against mobility both intellectually and emotionally.' Hess was both geologist and increasingly mobilist. The confrontation was to emerge in the open after Hess had published his tensional explanation of the mid-Atlantic ridge.

Ewing was disdainful of Hess's model, and at the First International Oceanographic Conference in 1959[16] persisted in maintaining that the ridge would be found to be capped by folded limestones, as if it were some range of hills in Texas. By 1954 Ewing had obtained a massive quantity of gravity data from the US Navy submarines which was analysed by a former colleague from Lehigh, J. Lamar Worzel. Joe

Worzel was 'a true believer who actually worshipped Ewing intellectually'. Together, they published a refutation of Vening-Meinesz and
Hess; the deep-ocean trenches were not formed by a pulling down and
squeezing together of the ocean crust but were deep tensional
features.[17] The rift between Hess and Ewing was growing wider and
deeper; Ewing's opinion of Hess found its mirror in Hess's of Ewing:
'No one has collected more data and contributed less ideas towards the
study of ocean basins than Maurice'.[6] Yet despite the strength of his
fixist conception of a physicists' solid Earth, Ewing sustained sufficient
impartiality to tolerate at Lamont dissident opinion on the condition
that it shared his enthusiasms.

The Lamont dissidents were to be influential. The first of them
arrived in 1952 from the University of Iowa where he had become a
convert from Palaeontology to Marine Geophysics after hearing
Ewing lecture. Bruce Heezen was set the task of creating a map of the
Atlantic from all the existing soundings. The data were insufficient to
contour the topography; the subsurface was still the preserve of the
military. At the height of the Cold War, from 1952 until 1962, the US
Navy classified all deep-ocean soundings. In 1957 Heezen with his
assistant Marie Tharp chose to produce an illustration as if the ocean
floor were a whole new world on some distant planet; the ocean waters
become invisible.[18] The featureless abyssal plain was far flatter and far
more extensive than any such feature on land. And the mid-ocean
ridge was as wide and as high above the surrounding floor, as the
range of mountains that rise out of the mid-western plains in
Colorado. The decision to produce an illustration rather than a map
allowed the new landscape to have immediate impact. By 1955 Heezen
believed that the mid-ocean ridge was a tensional feature.

It was not until the mid 1950s that the axial rift, first noticed in the
Carlsberg Ridge of the Indian Ocean, was recognised (first by Maurice
Hill at Cambridge and then by Heezen and Ewing) as lying at the
centre of the mid-Atlantic mountains throughout their length. The
mid-Atlantic rift was at the heart of Heezen's contemplations when
in 1958 he read the proceedings of Carey's Hobart Symposium on
continental drift. If basaltic magma was being inserted into the rift,
the rift itself should be widening. From 1958 Heezen became an
expansionist: Widening at the ridge meant that the whole North
Atlantic was growing. Heezen found that his own discoveries
provided direct confirmation of an inflating planet.

Sea-floor spreading

In 1956 Ewing had maintained the geologist's belief: that a deep drill
core in the oceans would pass through sediments that could offer a
complete history of the Earth. From this time on he refrained from the
furious arguments that were beginning to break out, preferring to
publish papers innocent of any greater scheme, and to present his

opinions only privately in discussion. In his lecture on 'The shape and structure of the ocean basins' delivered at the first International Oceanographic Conference in New York, 1959, he gave no intimation of any broader scale of activity. This did not pass unnoticed by the illustrious audience, which included all the most significant participants in the subsequent discoveries: Bullard, Hess, Wilson, Drummond Matthews from Cambridge, Henry Menard from Scripps, Robert Dietz, then a naval oceanographer, and Bruce Heezen.

When Ewing did explain his ideas it was as if he was attempting to make a compromise whereby his own marine researchers would in no way interfere with the continents and the beliefs that belonged to the geologists, such as Bucher. Any convection was restricted only to the ocean floor; the continents were immobile; the movement of the oceans compressed the edges of the continents piling up mountains. Marine geophysics would move tip-toe around the science of Geology without awakening it. However this liberalism towards a dynamic Earth, demonstrated when the ideas were still young, was to disappear in the next few years as the resurgence of mobilist views demanded a sterner denunciation.

On Hess's invitation Bruce Heezen had presented his theories in a number of talks delivered at Guyot Hall in Princeton. While Heezen's conception of the Earth focused on the mid-ocean ridges, Hess had begun his research investigating the other major topographic features of the oceans: the trenches. Although he had conceived of the importance of convection, this was on a scale that could explain major topographic features but not the overall configuration of the continents. Under Heezen's influence Hess began to contemplate convection beneath the oceans on the scale of the Earth. The ideas were pulled together and written up as a draft chapter for inclusion in volume 3 of *The Sea*, being edited by Maurice Hill at Cambridge. The essay entitled 'The origin of the ocean basins' was subsequently withdrawn from the book, but not before it had been circulated among a number of students. In 1960 Hess revised certain details of his manuscript and had it privately distributed as a report of the Office of Naval Research. Two years later it was formally published.[19] Heezen's geo-artistry of the Atlantic floor was now to inspire a literary imagination.

'I shall consider this paper an essay in geopoetry' began Hess unconfidently. Following a scheme originally proposed by Vening-Meinesz, after planetary condensation a single planetary convection cell had piled up continental crust only in one hemisphere. Once the core had developed, the convection became subdivided and restricted to the mantle. The smaller convection cells fragmented the original continent, and continue to separate the pieces today. The demonstration of all this lay in the oceans. The mid-ocean ridges lie above rising convection currents. Below 500°C the mantle peridotite becomes converted through contact with heated waters, into the serpentinite that Hess believed was the normal material of the ocean crust. Around

Rothé's 1953 map of earthquakes of the Atlantic and Indian Oceans (with permission from the Royal Society of London)

some oceans, in particular the Pacific, the ocean crust descends back down into the mantle, the serpentinite becoming cooked back to peridotite, expelling the water into the oceans; 'the great advantage of serpentine – it is disposable'. Sediments that have accumulated on the sea floor become squeezed and metamorphosed in the 'jaw-crusher' between ocean floor and continent and so become added on to the continents. Hess's convection, unlike that suggested by Ewing involved the continents, moving them as passive wanderers on the roofs of the convection currents. Hess had finally come up with a theory closely similar to that first proposed 30 years previously by Holmes. The geopoetry had a rhyme.

In 1961, before Hess had achieved publication, Robert Dietz, working at the US Coast and Geodetic Survey produced an adaptation of Hess's ideas,[20] with some significant variations; the peridotite to serpentinite transformation became altered to one between the extreme high pressure form of basalt, known as eclogite, and its volcanic analogue. More significantly he found a name for Hess's geopoetry: 'sea-floor spreading'.

During the early 1950s a French seismologist, Jean Pierre Rothé, had analysed the information collected by the International Seismological Centre at Strasbourg (transferred to France at the conclusion of the First World War, when France had captured a section of the German Geophysics research initiative). At the 1953 Royal Society meeting Rothé produced a map of earthquake epicentres principally for the Atlantic and Indian Oceans.[14] He showed that the earthquakes associated with the Carlsberg and mid-Atlantic ridges were part of a single feature that could be traced from the East African rift valley, out through the Gulf of Aden into the Indian Ocean and on round the southern tip of Africa, half-way between that continent and Antarctica.

In 1956, immediately prior to the International Geophysical Year, Ewing, in collaboration with Bruce Heezen, made the prediction that a rifted ridge would be found wherever the 34 000 mile (54 400 km) narrow band of earthquake epicentres could be traced through the centres of the oceans.[21] The existence of this, 'the longest mountain range on the planet', was confirmed over the next three years wherever the line of the earthquakes was intersected. In 1958 Henry Menard, a marine geologist at Scripps, decided to test how closely these ridges followed the ocean centres. He found that 'the median line follows the sinuous crest of the oceanic rises and ridges through the Atlantic, Indian and Antarctic Oceans with remarkable consistency.'[22] This uncanny symmetry could only mean that the ridges were related to the positions of the continents. No other feature on Earth behaved according to such simple rules over such vast distances. Geometry was beginning to reassert itself in the configuration of the planet.

One of the most surprising results to emerge from the deep-ocean Geophysics concerned heat-flow. Just as the continental heat-flow researchers, Daly, Joly and Holmes, had been led by their studies into mobilist views, so now a new generation was inculcated into a mobile mantle from the evidence of the ocean-crust temperature gradient. The first successful measurements were made in 1947 and 1948 by the Swedish Deep Sea Expedition using an unreliable 11m geother- mometer that provided only two good readings, both of them strangely high.[23] Two years later, Edward Bullard, while working at Scripps in California, helped to design and build (with Roger Revelle and Arthur Maxwell) a shorter, stronger heat probe, tested in the Pacific in 1950 and the Atlantic in 1952. The readings were also much hotter than anticipated. Careful analyses were made of the sea-bottom sediments to test for previously unrecognised radioactive materials but these were unremarkable; 'perhaps' wrote Bullard 'heat is being transported outward by convection in the mantle'.[24]

Bullard persisted with his heat-flow determinations and by 1956 had discovered that the heat-flow on the mid-ocean ridge was five or six times that found over much of the rest of the oceans.[25] If there was convection, then the ridges must lie above the main thermal upwelling. Bullard had returned to Britain in 1950 to take up the Directorship of the National Physical Laboratory, a job with some of the breadth of his wartime role. In 1957 he became once again head of the Cambridge Department of Geodesy and Geophysics, a position only honoured with a professorial chair in 1964. Yet within Bullard's eclectic range of geophysical research interests there was a significant gap.

While clues to ocean origin were appearing throughout the new marine geophysics, the most remarkable evidence emerged from the post-war study of marine geomagnetism. The technical development that made the study possible came out of the wartime research programme into finding some means to detect submerged submarines from the air.

Marine magnetics

The original Second World War fluxgate magnetometers had been developed by an American physicist, Victor Vacquier, working for Gulf, to provide a continuous magnetic reading for an airborne survey; a sudden change in the magnetic field over the sea could reveal to the pilot the presence of a submerged submarine. The fluxgate magnetometer suffered from problems of drift and required fairly frequent correlation and adjustment. It was also cumbersome and relatively fragile. However, by 1952 Ewing had succeeded in adapting such an instrument to be towed behind a ship.

In 1951 as Ronald G. Mason completed his Ph.D. at Imperial College, London he was invited by Roger Revelle, the Director of the Scripps Oceanographic Institute in California, to work on marine magnetic data. Mason managed to borrow from Ewing a fluxgate magnetometer to be towed behind the vessel *Horizon* on the Scripps, Capricorn Expedition of 1952. Accompanying him on this voyage was an electronic technician, Arthur D. Raff, who became in the following years Mason's special assistant.

In 1954 Scripps was invited to test a new more accurate type of drift-free instrument, the proton-precession magnetometer, which offered an accurate calibration for the readings. With the measuring techniques now mastered, in 1955 Mason learnt that the US Coast and Geodetic Survey had been commissioned by the Navy to undertake a detailed hydrographic survey from the beach to the ocean floor within 300 miles (500 km) of the western coast of North America, in order to gather information needed for the protection of the new nuclear submarines. The Navy readily agreed to the inclusion of a magnetic survey.

The cruise had already been underway for two months before Mason and Raff got their equipment on board. The ship was running a grid pattern of lines spaced at 5 mile (8 km) intervals and had started the survey in the south off Baja California. By the time Mason had begun his measurements they had arrived off San Diego. The survey lines were running E–W and Mason processed and plotted the information as it was collected. These first results were extraordinary; land-based aeromagnetic maps were always complex patterns with only abstract intimations of order, yet off the San Diego coast the magnetic anomalies all lay parallel to one another with a simple north–south orientation, as though this was a floorboard arrangement made of planks (15–25 km) wide and hundreds of km long.

Mason was overcome with caution. His expectations were only that the ocean magnetics would provide new information; not that some pattern would immediately present itself that showed a greater degree of regularity than that possessed by any other feature on the Earth's surface. At first he had doubted whether anyone would believe his results. He had been a member of the Imperial College faculty since 1947 and now had considerable teaching commitments. He very nearly abandoned the Pacific survey programme but was prevailed upon to continue by Roger Revelle. The programme was eventually salvaged by Raff who continued to work on the *Pioneer* after Mason had returned to England and after admonishments from Raff's superiors to return to the workbench at the Marine Geophysical Laboratory at Scripps. Raff sent back the data as they were collected to Mason in London. Mason calculated the size of the anomalies, and meticulously contoured them. As contours they were hard to read, but when either positive or negative anomalies were blacked out, the pattern was that of zebra stripes.

After various preliminary charts had been circulated around Scripps, Mason and Raff published their first map of the southern part of the region in 1958[26] and had completed a magnetic anomaly map for the whole coastal Pacific belt by 1961.[27] The anomalies did not all run north–south; if the arrangement was that of planking, then the Pacific Ocean floor was constructed out of some elaborate form of marquetry.

H. William Menard of Scripps had been supervising some of the

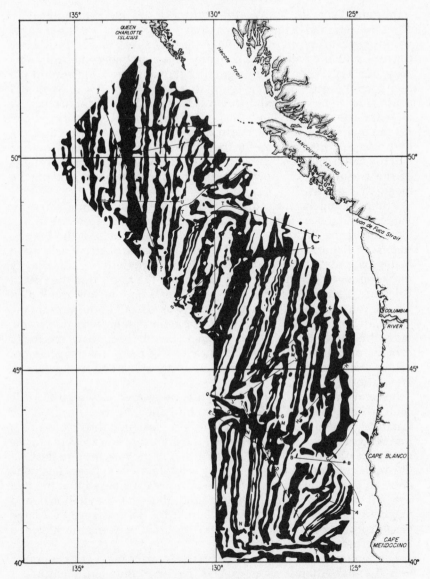

Raff and Mason's 1961 map of the total magnetic field anomalies south-west of Vancouver Island (with the permission of the Geological Society of America)

Institute's own hydrographic work, attempting to map the Pacific in parallel to Heezen's Atlantic programme. In 1955[28] he had discovered several bands of massively disturbed topography running due west from the Californian Coast for hundreds of kilometres into the Pacific. The two most significant features he had labelled the Mendocino and Murray fracture zones. These fracture zones provided the most important discontinuities in the pattern of the magnetic anomalies. After Mason had produced his first map, Victor Vacquier, now working at Scripps on marine magnetism, realised that at the Murray fracture zone the pattern of anomalies had merely become shifted by some 135 km. When more magnetic data were available around the Mendocino fracture zone, he found that if the anomalies to north and south were to be matched, the two sides must have become offset by an extraordinary 1120 km.[29] Such displacements seemed incredible, yet the magnetics gave the ocean floor a new texture. It was as if the hydrographic information for the oceans produced a murky outline and the magnetics a picture in vivid colour. Any theory for the origins of the oceans would now first and foremost have to explain the magnetic patterns.

While Ewing had captured much of the mid-Atlantic ridge for Lamont, and the eastern Pacific was studied out of Scripps, the Indian Ocean and the Red Sea were, as if by the traditions of the Empire traders, still the sanction of the British. The east African rift valley that had provided the site of Bullard's doctoral dissertation on gravity measurements, led into the Red Sea and to the north into the rifts of Palestine. East of Aden the Carlsberg Ridge ran out into the Indian Ocean. Whereas the mid-Atlantic ridge was lost out in the midst of the ocean, in the scorched deserts of Ethiopia the oceanic structures came onto the land. Bullard, like Holmes before him, had originally considered that the east African rift valley was the site of compression. Along the line of the valley there was a marked reduction in gravitational pull.

It seemed natural that these geophysical properties would stay the same where the rift valley passed into the Red Sea. In 1957 the Cambridge geophysicist, Ronald Girdler, presented the results of his Red Sea gravity survey that showed the opposite: an excess of gravity running along the axis of the rift.[30] Further to the east, around the Gulf of Aden, an aeromagnetic survey had located a large magnetic anomaly along the centre of the sea. Girdler considered that the Red Sea rift was a tensional feature, and that the gravitational and magnetic anomalies could be explained by the presence of a large basaltic intrusion of solidified magma inserted between the stretched lighter crust of the continents.

The Lamont group had been the first to describe magnetic anomalies over the mid-Atlantic ridge in 1953. By 1957[31] they had measured a magnetic high along all the length of the central rift. The discovery of the continuity of the mid-ocean ridges as well as the sudden renaissance of theories of continental drift and the expanding earth placed the rifts at the focus of all Marine Geophysics. By 1959 the global rift system seemed everywhere to be a tensional feature that was filled with intruded basaltic rock of high magnetic susceptibility. In 1959 Hess had sent his geopoetry to Cambridge.

To combine the rifts, the zebra-striped Pacific and the geopoetry needed the results of a new research programme that had nothing to do with the scope of Marine Geophysics. Time and rocks; these were the ingredients both of Geology and of the study of magnetic reversals. That certain rocks were magnetised contrary to the present-day magnetic field had been discovered in the 19th century. After Pierre Curie had shown in 1895 how remanent magnetism became imprisoned in certain minerals as a rock cooled from high temperatures, it became possible to undertake a more systematic study of the magnetism of past lava flows. The first successful attempt to peg such reversals onto the geological timescale was made by Motonori Matanuyama of Kyoto Imperial University, who worked for several years from 1926 on basalt specimens from Japan and Manchuria.[32] The polarity of the rock's remanent magnetism appeared to be dependent on the rock's position within the recent geological record. During the beginnings of the Pleistocene epoch, the magnetic field seemed to have been reversed.

The whole phenomenon of magnetism was sufficiently perplexing to leave unresolved the question of self-reversal. If rocks could spontaneously preserve a counter-magnetic field, then it would threaten any regular timescale with anarchy. Self-reversal provided extremely interesting research for a physicist; but only global reversal offered a new key to Earth history. There was no immediate way in which the two phenomena could be resolved. Instead it was necessary to undertake an arduous research programme, collecting and dating suitable lavas until the weight of evidence for the corroboration of the age of reversals would allow all exceptions to be discarded.

Through the 1950s in Iceland and the Soviet Union a number of attempts were made to construct such a reversal timescale:[33, 34] By 1960 the case for global reversals was complete, apart from the increasing evidence that self-reversal could and had occurred.

The state of uncertainty could never be resolved as long as the reversals found in volcanic lavas were being correlated with a geological timescale; a timescale of irregular epochs, whose boundaries were defined from the fossils found within neighbouring sediments. In place of stratigraphy, of talk of the boundary between the Pleistocene and Pliocene, of Neogene, and Miocene, a new numerical timescale had to be erected.

Radio-isotope dating measures the concentrations of the daughter products of radioactive decay relative to the parent atoms. Rock dating utilises the pristine radioactive isotopes, isotopes that were created in the nuclear furnace of a star before the Earth was born. To have survived since the origin of the Earth these isotopes must have a sluggish rate of radioactive decay. For rocks that are geologically very young, less than one-thousandth the ($4\frac{1}{2}$ billion year) age of the Earth, the amount of daughter products formed in such decay is therefore extremely small. Thus the science of young rock dating had to develop equipment of extreme sensitivity.

The discovery that argon (^{40}Ar) was a decay product of potassium (^{40}K) was made immediately before the war. The wartime Manhattan

Project, to build an A-bomb had close associations with the University of California at Berkeley and this contact ensured a continuity of post-war research into a wide range of problems in Nuclear Physics. From 1956 when the physics of the potassium–argon system had been fully comprehended, it was left to Berkeley to develop the dating technique for younger geological samples through innovations in the whole procedure for extracting the argon gas.

By 1960 potassium–argon dating had closed the wide age gap between the pristine isotopes and carbon 14 (which utilised the decay of a relatively short-lived isotope formed continuously in the upper atmosphere by cosmic ray bombardment). Berkeley's lead in the techniques of young rock dating was about to be narrowed through the setting up, with the assistance of Berkeley researchers, of similar laboratories at the Australian National University in Canberra and at the US Geological Survey in Menlo Park, California.

However, the first attempt to incorporate radio-isotope age dates into a reversal timescale was made by a Dutch professor of Stratigraphy and Palaeontology.[35] Martin Rutten had constructed his scale largely from the older geological stratigraphic data but pegged it on three preliminary age dates taken from the Berkeley researchers. At Berkeley during the late 1950s, two geomagnetists, Richard Doell and Allan Cox, had been sited in a hut next door to the young-rock dating laboratory. By 1962 they had begun to undertake the first experiments specifically designed to correlate age dating with magnetic pole reversals. On 15 June 1963, having moved to Menlo Park, they published, in collaboration with a younger Berkeley researcher G. Brent Dalrymple, the first radiometric reversal timescale.[36] Ten days later they had been awarded $120000 by the National Science Foundation to further their research. It was only a matter of time, travel and energetic work for the scale to become finalised.

The mid-ocean tape recorder

At the most formative first International Oceanographic Congress in 1959, Bullard had provided a review lecture in which he stated that the palaeomagnetic evidence provided 'a strong case for relative movements of the continents', and that 'the mid-Atlantic ridge and particularly its central valley might be regarded as the place where the Atlantic is at present widening'.[16] Bullard remains sceptical, but yet in these statements it is clear that his conversion to drift is underway. The English geomagnetists, in particular Runcorn, had found convincing evidence for drift in the different polar wandering paths between the continents. Bullard had already seen the implications for the oceans.

That magnetic field reversals imprinted in the crust of the sea floor could be responsible for some of the curious marine magnetic anomalies was first suggested in two separate articles published by

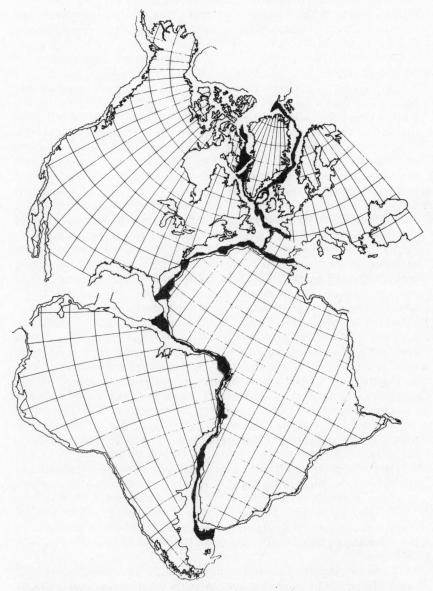

Bullard's (with O. E. Everett and A. G. Smith) 1964 computer reconstruction of the Atlantic Ocean margins, mapped at the 1000 m depth contour Areas of overlap have light shading, and gaps are black (with the permission of the Royal Society of London) (Ref 9, Chapter 7)

English researchers in 1960. The Cambridge geomagnetist Ronald Girdler had written a paper in 1959 with George Peter proposing reversals in the rocks beneath the Gulf of Aden. After the journal *Geophysics* refused the paper publication, it eventually appeared 'in a much watered down version'.[37] The second paper by Maurice Hill of Cambridge with A. S. Laughton and Thomas D. Allan employed reverse polarisation to explain a magnetic anomaly on a sea-mount 250 km north of Madeira.[38] None of the authors of these papers intimated that reversals might be fundamental to the magnetic pattern of the ocean floor.

Dietz's 1961 exposition of sea-floor spreading had incorporated Mason and Raff's eastern Pacific magnetic anomaly pattern as information consistent with some kind of growth of ocean crust out from the ridges. The stripes themselves that developed perpendicular to the direction of spreading were, however, attributed to some configuration of linear crustal stresses. The great fracture zones could mark shears bounding sections of ocean crust spreading at different speeds. Dietz had repeated his seafloor spreading theories in a 1962 book entitled *Continental drift* (edited by Runcorn), intended to commemorate the half century since Wegener published his first article 'On the origin of the continents'. The book contained chapters by Heezen, still advocating an expanding earth, and Vening-Meinesz, contemplating his episodic thermal convection. Even Victor Vacquier in an expansive mood appeared to be advocating crustal mobilism; claiming that as the marine magnetic anomalies did not continue onto the continents so the oceanic crust must be regenerating at different rates.

In 1960 on the maiden voyage of the research vessel *Argo*, Vacquier had visited Carey in Hobart and left clutching a copy of the 1956 Hobart Symposium proceedings. (Carey's visitors were mostly those involved in round-the-world cruises, but at least this offered him access to some of the crucial marine geophysicists.) Even in 1962 Vacquier's earlier enthusiasm for Carey's ideas has persisted; they are mentioned as one possible explanation of oceanic crustal regeneration, another being the rise of mantle material along the mid-ocean rift. Vacquier shows remarkable prescience in stating that 'The . . . lineation of the magnetic anomalies would be the record of the regenerative process. Therefore a band of strong magnetic lineations should run parallel to the great oceanic rises'; remarks that subsequently he was to claim to have forgotten and to have been 'a low-probability postulate'. In the lull before the storm, before critical censure has polarised an argument, so opinions may be unfettered. In the company he kept, amidst the pages of this book, he could flirt with mobilism.

The predominant opinion was that the magnetic anomalies reflected some difference in magnetic susceptibility between the crustal rocks. The pattern required that these rocks must be distributed as plank-shaped bodies with deep, sharp, lateral boundaries – known in Geology as dykes. Yet almost nothing was known about the actual magnetic properties of the sea floor. In 1961 having completed his Cambridge Ph.D. on 'Rocks from the eastern North Atlantic', Drummond Matthews was appointed to a junior teaching post and instructed by Maurice Hill to supervise the British research effort on the International Indian Ocean Expedition. Maurice Hill had persuaded the British Navy to lend a ship, HMS *Owen*, on which Matthews spent six months at sea.

Recognising that the late-arriving American contingent would never be bettered in mapping a large area, Matthews chose to survey at least one square straddling the Carlsberg Ridge in greater detail than had ever been achieved for any section of the ocean. With the gravity, bathymetric and magnetic data for this patch as well as a more general coverage of a much wider region, he returned to Cambridge in November 1962 and found he had been presented with a research student to supervise: Fred Vine.

The Cambridge Department had sustained research in geomagnetism throughout the 1950s. At the beginning of the decade Edward Irving and Jan Hospers had completed Ph.D.s on magnetic reversals; subsequently the topics had changed to polar wandering and drift. The department, like Lamont, was in a converted country house, but a far humbler and less pretentious affair, little more than a country parsonage. The intellectual effort suited this pleasant, airy, English simplicity. Vine and Matthews worked in the coachman's quarters above the stables where the marine sediment cores were located. Vine was given the task of discovering, through a series of computed structures, the nature of the formations that gave rise to the magnetic anomalies.

In January 1962 Harry Hess had come to Cambridge at Bullard's invitation, to be the guest speaker at the student-run Inter-University Geological Congress, titled that year: 'The evolution of the North Atlantic'. Hess outlined his geopoetry, still awaiting publication, and later that spring Vine presented a talk on Hess's ideas to the undergraduate geology society, 'The Sedgwick Club'. Having assimilated Hess, been taught structural Geology by an enthusiastic drifter (Brian Harland), and now changed to an out-of-town research department headed by the most exuberant and powerful geophysicist in Britain, Vine was in a perfect milieu for a breakthrough.

On the other side of the Atlantic the indirect influence of the other of Field's pupils, J. T. Wilson, was about to make itself known. Lawrence Morley had learnt palaeomagnetic techniques under Wilson in 1948 at Toronto and had gone on to complete a Ph.D. (supervised by Wilson) during the period when Bullard had held a Physics professorship. Toronto was almost unique for the alliance of Geology and Physics that it could offer students, an enlightened condominium that transcended all the 19th-century boundaries to the disciplines.

Like Wilson, Morley had carried out extensive aeromagnetic surveys and by 1963 was in charge of the Geophysics Division of the Canadian Geological Survey. Frustrated by the slow speed of ship-board magnetometer surveys, Morley had chosen to survey the Newfound-land continental shelf from the air. He had become fascinated by the results of the Mason and Raff survey of the eastern Pacific and eight months after it was first presented in August 1961 read Robert Dietz's published account of sea-floor spreading that had been based on Hess's geopoetry. These two papers became as one. Following the premature results of the Icelandic studies on magnetic reversals, Morley located all the rock magnetism in the top few kilometres of the ocean crust where the rock was cooler than the Curie point, and calculated that the striped pattern of anomalies could be explained with a reversal every million years if the convection current that drives the sea-floor spreading moves at approximately 3.5 cm per year. Morley fortified this suggestion from other lines of evidence: first, the ocean crust seemed to be younger than 140 million years; second, many guyots have curiously suppressed anomalies, suggesting that their constitutent volcanic rocks have a magnetic field contrary to that of the Earth. The synthesis was sent off as a short paper to *Nature* in February 1963.

Two months later Morley was informed that *Nature* 'did not have room to print' his contribution so, unabashed, he immediately dispatched the letter to the *Journal of Geophysical Research*. The editor replied in late September, confessing that the delay had been caused by his absence during the vacation, that the paper had meanwhile been turned down by a referee, and that 'such speculation makes interesting talk at cocktail parties, but is not the sort of thing that ought to be published under serious scientific aegis.' The summer vacation had, however, seen a sea change; on 7 September 1963 Vine and Matthews had managed to get their own version of cocktail party chatter published in *Nature*.[39]

Vine had arrived at the theory by a simple series of stepping-stone ideas. He had tested, in the manner of Hill before him, the possibility that the difference between the magnetic anomaly found over two sea-mounts in Matthews' detailed survey of the Carlsberg Ridge could be explained through one having normal magnetisation, the other reversed. Having demonstrated that reverse magnetism was present in the ocean floor, he was led to incorporate it into his enthusiasm for Hess's spreading ocean crust. The orthodox view on ocean magnetic anomalies was that they represented some variation in the mineralogy of the rocks of the sea floor. In Vine's elegant explanation, the ocean crust could be uniform but merely generated during differing periods of normal and reverse magnetism.

Vine wrote up the idea, showed it to his research superviser who had been honeymooning at the moment of enlightenment, and then to the head of the Marine Geophysics Programme, Maurice Hill, who 'just looked at me (Vine) and went on to talk about something else'. Hill was a physicist; Vine had entered Geophysics after taking courses in Geology, and specialising in his final year in Petrology. The identification with Hess was heightened by their both having

Fred Vine
British geophysicist

followed this mental path, from igneous rocks to Geophysics, in the study of the sea floor. Bullard was more encouraging but disavowed his name on the paper, and so Matthews, whose data provided the substance of the discussion, gave the original note greater scientific solidity, thereby, as Vine admitted, expediting publication. The magnetic patterns that the group had gathered from the Indian and Atlantic Oceans were still very noisy and Vine made no suggestion that such anomalies might actually lie symmetrical about the mid-ocean ridge.

Although Morley and Vine gained the same intimation, only Vine was undertaking active research and, more significantly, only Vine could actually apply the idea in the interpretation of primary data. By 1963 it was inevitable that someone should have put the story together. Following his humiliation, Morley eventually presented his ideas in a short talk delivered to the Royal Society of Canada, meeting in Quebec City in June 1964, which was published later that year.[40]

The publication of the Vine and Matthews note provoked hardly a ripple. In the knowledge that his enthusiasm for the spreading sea floor had preceded the attempt to break the code of magnetic anomalies, even Vine himself was doubtful of the validity of his own theory. Yet there was a far greater momentum to the discoveries. The dark horse among Field's geophysicists, J. T. Wilson, after taking no interest in marine investigation became concerned with all manner of global geophysical problems during the International Geophysical 'Year' of 1957–9 and in 1959 through contact with the English palaeomagnetists including Irving, and Runcorn began to be converted. In 1960 he was privileged to read a draft of Hess's geopoetry and on return from a National Academy of Sciences trip to Antarctica stopped over in Hawaii where he visited the Upper Atmosphere Observatory 3600 m up on Mauna Loa.

The volcanoes of the Hawaii islands were known to have passed through a life cycle in which progressive changes in the composition and quantity of the ejected magma altered even their overall morphology. Each island and each volcano showed the same changes but the islands to the north-west were closer to the end of the cycle; those even further from the largest island of Hawaii were now all extinct and rapidly becoming overwhelmed by the sea. Wilson thought that the islands could lie atop a massive stationary thermal plume in the underlying mantle that provided a line of volcanoes getting younger to the south-east as the sea floor drifted overhead; a factory chimney sending billows of smoke into the breeze. He wrote up the idea and sent it to the *Journal of Geophysical Research* which refused publication. Blacking of ideas by innovative Canadian geophysicists seems to have become commonplace. Wilson transferred the paper to the home-based *Canadian Journal of Physics* where it appeared in 1963.[41]

The simplicity of this idea of age-graded lines of oceanic volcanoes led him to search for other examples. While in no position to organise marine research, he had become hungry for information and from 1960 began collecting all published data on oceanic islands. In 1963, under a grant from the US Air Force, he wrote up two volumes of notes on their Geology and Geophysics (notes that were never published, but copies of which were sent to certain institutes). Islands, he reasoned, provided easy access to details of the hidden oceans. Once those such as the Falklands and Seychelles, built out of continental crust, had been excluded, some simple stories emerged. As one travels further from the mid-ocean ridges age dates from the volcanic lavas prove the islands to be progressively older. The islands themselves tend to lie along straight lines and where they lie on the mid-ocean ridge, they form the intersection between two such lines. Without a research ship, simply through exploration in the geological libraries, Wilson had shown that the rate of movement necessary to carry 36 million year old rocks on Bermuda away from the mid-Atlantic ridge was about 3 or 4 cm per year.

7
The Big Bang

The rebirth of Seismology

The first American hydrogen bomb, code-named 'Mike', devastated the central Pacific island of Elugelab, on 31 October 1952. The 10 megatonne explosion had required a hydrogen refrigeration plant the size of a house to condense the deuterium fuel prior to detonation, and it was not until 1 March 1954 that the trick of using the solid compound lithium 6 deuteride allowed the construction of a deployable weapon. This second explosion, code-named 'Bravo', became the turning point in the world's acceptance of the bomb. The coral reef on Bikini Atoll, which had been blasted into fragments by the surface explosion, began to fall as a light rain of highly radioactive fallout over a region of 18 000 km² of the Pacific. After almost a decade of neglect, intimations of the terrible effects of radiation precipitated an international outcry about the new, more fearful, destructive powers of the hydrogen bomb. From the two-piece bathing suit to the paperback best-sellers, the fear of nuclear holocaust came to dominate the culture.

In 1892 Father Frederick L. Odenbach, was assigned to the Jesuit College of St Ignatius in Cleveland, Ohio, where in 1896 he started a meteorological observation station. In 1900 he built his first seismograph, installed at the College. On 2 February 1909 Odenbach addressed a letter to all the Jesuit Colleges in the United States and Canada inviting them to participate in an experiment. 'With a small outlay at a number of our colleges we would be in a position to do the *great thing in seismology.*'[1] The Jesuits and the Roman Catholic Church were about to become wedded to the science of global Seismology.

The letter was successful in encouraging the installation of 18 instruments around North America. However, during the First World War the network became fragmented and run down, to be reinvigorated in 1925 by a mathematical seismologist Father Macelwane, newly appointed as Professor of Geophysics at St Louis, Missouri. The Jesuit Seismological Association received support from William Bowie of the US Coast and Geodetic Survey and became co-ordinated through

Macelwane's own St Louis department. The seismological stations that had already been set up at Jesuit Colleges in Cuba, Colombia, Spain, Hungary, Lebanon, Bolivia, Philippines, Madagascar, China, England and Australia chose to maintain their own autonomy and offer 'informal co-operation', because it would be 'less open to misunderstanding'; the misunderstanding presumably being that of a global Jesuitical conspiracy. However, they too shared the sense of promise and destiny. The Director of the Granada Station wrote that the collaboration 'would allow us to do research of absolutely first order with a sum of money that would not suffice to mount the smallest astronomical observatory'. Through the perseverance of individual Jesuit seismologists, the new global Seismology could traverse national frontiers, locating and measuring earthquakes even into countries such as Switzerland and Norway from which Jesuits were excluded.

The Jesuit network was the third (but most durable) attempt to create a global seismology. The 27 year old German astrophysicist Ernst von Rebeur-Paschwitz had inadvertently (on equipment designed to measure lunar gravity), in the spring of 1889, detected ground movements at Potsdam and Willemshaven from a major earthquake in Japan. He worked to detect large distant earthquakes for the next few years and, while already bed-ridden from a terminal illness, in 1893 proposed that a global network of stations could encompass the world's earthquakes. His vision began to be implemented after his death in 1895, by an English geologist and mechanical Engineer, John Milne.

After 1865 the desire for Western technology encouraged Japan to found new institutions and until 1900 to offer high salaries for '*oyatoi-gaikokujin*' (honourable foreign hirelings) to act as teachers. This attracted to Japan a number of bright young English scientists who found themselves in one of the most seismically active regions on earth. Milne became the Professor of Geology and Mining in the Imperial College of Engineering at Tokyo early in 1876. Following a moderate earthquake in Yokohama in 1880, Milne called a public meeting at which the Seismological Society of Japan was founded, and thereafter proceeded to develop a series of earthquake measuring instruments, culminating in the Milne instrument of 1892. Milne returned to England in 1895 to build an earthquake observatory in the Isle of Wight, that by 1899, through funding from the British Association for the Advancement of Science, was the hub of a global Empire network of 27 instruments.

In 1897 the German physicist Emil Wiechert had been appointed to the University of Gottingen where, in 1901, he inaugurated the first 'Institute of Geophysics'. In 1900, following a visit to an Italian earthquake, Wiechert developed his improved inverted pendulum seismograph. 'Wiecherts' were rapidly installed in centres throughout the German colonies, and were also used by the Jesuits. The momentum was with the Germans and in 1903 it was they who obtained the headquarters of the International Seismological Association at the geophysical heartland (the home of the Beiträge die Geophysik) in Strasburg.

Both the British and German global networks fell apart during the First World War. German global ambitions ended with the confiscation of their colonies and the British network lost its purpose with the death of Milne in 1913. The global stations themselves were not inviolate: the British Cocos islands instrument was destroyed by a landing party from the German cruiser *Emden* soon after the outbreak of hostilities. No longer enriched by Japanese teaching contracts, British seismology was already stagnating; in contrast to Germany there were no geophysical institutes to train Milne's successors.

An attempt was made in 1910 to create an American Bureau of Seismology, but was rejected by Congress, partly because publicising earthquakes damaged national morale. Thus it was the Jesuits who were licensed and later encouraged to act as the vicarious co-ordinators of American global seismology.

In 1954 Father Rheinberger, the reader of records at the Observatory at Riverview College, Sydney, Australia, noticed a tiny deflection on the seismogram which could correspond with the description given in the news about the explosion at Bikini. The Director of the Observatory initiated a search through the records from other Jesuit stations and soon found evidence of four such hydrogen bomb tests. For the first time, man-made explosions were equivalent in energy to a big earthquake and could be detected all around the globe. The New Zealand seismologist Keith Bullen (who had gained a Ph.D. at Cambridge under Sir Harold Jeffreys before becoming Professor of Applied Mathematics at Sydney) was invited to work on these records.

Mathematical Seismology, as it existed at the beginning of the 1950s, was almost entirely concerned with the internal structure of the Earth as revealed by the passage of seismic waves; the Seismology of Jeffreys. Earthquakes were considered useful generators of a seismic impulse but had many drawbacks, occurring at unaccountable depths, at unpredictable times. Through some simple assumptions about distance from Bikini, Bullen found that the bombs were always fired close to 5 minutes past the hour. Knowledge of the exact location and the firing time made possible a more detailed analysis of the passage of the seismic waves through the Earth. In 1955 Bullen had suggested that one or more fission bombs should be employed for seismological research into the structure of the Earth's core. At the International Association of Seismology and Physics of the Interior of the Earth at Toronto in 1957, Bullen delivered the Presidential address, claiming more forthrightly that 'The evolution of nuclear knowledge now offers the possibility of revolutionary new attacks on important problems in seismology.'[2] The proposals were widely condemned; seismologists never got their bomb, but through Bullen's work, Seismology had proved itself capable of contributing to the nuclear age.

In January 1958 Linus Pauling, Chemistry Nobel Prize winner, presented a petition to the UN signed by 11000 scientists from 49 countries, calling for an international test ban. The weight of public pressure finally encouraged Eisenhower to make overtures to Chairman Bulganin and Premier Krushchev and technical discussions preliminary to a complete test-ban treaty commenced in Geneva in July 1958. Amidst the nuclear physicists attending the talks there were geophysicists, such as Edward Bullard, as well as a new breed of post-war seismologists, of whom the most impressive was Frank Press, the chairman of the US delegation, who had received his Ph.D. at Lamont. How could such a complete test-ban treaty be guaranteed? Windblown radioactive fission products and the shock wave of the blast itself promised no problems for detecting above-ground tests, but a programme of detection for underground explosions had yet to be attempted. While the Soviet Union was optimistic about the policing of a treaty, the Western scientists became more and more concerned about the discrimination of small earthquakes from low-yield atomic bombs. An early calculation suggested that it would require 650 detection stations world-wide to offer a threshold of 1 kilotonne.

From being a tiny fringe group of scientists operating their seismographic stations on shoe-string budgets, in geographical isolation, the seismologists had been thrust into the blazing lights of international diplomacy. Dazzled by their new-found importance they had a naïve understanding of the bargaining in which they were to play such an important role. Optimistic claims were made for instrumental sensitivity, based on entirely traditional procedures: through independent, enthusiastic station directors, communicating with each other through the mail. As the seismologists were to discover, in practice, all methods were inadequate to identify every low-yield explosion correctly.

The opposition to the whole concept of a test ban found support from the calculations of Edmund Teller's colleague Albert Latter who showed that by setting off an explosion inside a large cavity it would be possible to cut down the amount of energy transmitted into the ground by a factor as great as 300. In late 1958, after the angry and contemptuous Soviet response to the shifting claims of the American scientists, Eisenhower's Special Assistant for Science and Technology constituted the 'Panel for Seismic Improvement', headed by Lloyd Berkner, whose brief was to investigate all ways in which the science and therefore the detection limits of Seismology could be improved.

The Berkner Report, published in June 1959, offered a total revision to the mode in which Seismology had previously been undertaken. Computers, communications theory, seismic source studies were all to be harnessed in a massive research effort. Berkner was most severe on the present shambling state of the science, and on the original claims made in 1958. A budget for two years of $53 million dollars was recommended, under the supervision of some Advisory Panel, itself closely connected with the National Academy of Sciences. The Bomb had caused an explosion in the science of Seismology.

Six months later the Advanced Research Projects of the Department of Defense set up the bomb-monitoring programme, Project Vela, that had three sections: Vela Hotel and Vela Sierra for the detection of high-altitude tests, and Vela Uniform, the seismological programme, for underground explosions. From 1960 up to 1971 nearly $250 million was poured into Seismology. Within this budget everybody benefited; large multi-arrays of instruments were set up in Montana, Norway and Alaska. All the previous seismological observatories gained extraordinary new research funding; geophysicists could collect large research grants (Wilson's famous 1965 paper was acknowledged as 'a contribution to the Vela Uniform programme'); and most significantly (alongside a covert system of stations operated around the periphery of the Soviet Union) from 1961, a World-Wide Standardized Seismograph Network (WWSSN) began to be implemented using the best instrumentation then available, set up at 120 sites around the world. There were some notable gaps in this coverage; the Soviet Union had declined to allow stations financed by the US Defense Department to be based on its soil.

After the 1962 Cuban Missile Crisis, both the USSR and the US made moves to break the deadlock in the on site inspection discussions; the US demanding a minimum of seven, the USSR a maximum of three. This diplomatic stalemate eventually required a compromise; the replacement of the Complete Test Ban with the Limited Test Ban that restricted all future tests to the underground. Seismology was no longer to act as the referee, but instead simply the observer, monitoring the enemy's bomb capabilities as each new device was tested. The Limited Test Ban Treaty was finally signed in Moscow on 5 August 1963.

By 1967 the WWSSN had spent $10 000 000 and was essentially complete. From 1968 the funding entered into a decline as the success of the monitoring programme led the US Congress to pull back on its full support. However, between 1961 and 1968 the WWSSN information, rapidly collated at the US Geological Survey in Golden, Colorado, became the most important scientific resource in the study of the Earth. For the first time a large sample of earthquakes were located with a reasonably even coverage, and their locations pinpointed with an accuracy that could correlate them directly with topographic and geological features. But above all the whole sample could now be studied, through the output of the computers, in a single global map. WWSSN and the computer had produced a whole-Earth science. The seismologists had now gained the equivalent of a 200" telescope that could be turned on those features already identified as significant by other geologists and geophysicists. Yet unlike these other sciences, Seismology observed activity continuing today, and not just the record of past processes. It was Seismology which was finally to reveal how the ancient, disordered, solid Earth was a living patterned planet.

Continental drift revisited

That most famous of all seismic events, the 1906 San Francisco earthquake, continued to be the Great American Earthquake, as much a part of American folk history as the Wild West and Hollywood. As Californian geology was mapped, the land to either side of the San Andreas Fault zone was found to be built from entirely different materials. The first attempts to match the two sides of the San Andreas were inhibited only by the implied magnitude of the displacement. In 1952 it was announced that there had been several hundred kilometres of movement.[3]

Just as the 1880s and 1890s were the age of overthrusts and nappes, the post-war period became the era of 'transcurrent' faults that revealed massive horizontal displacements. In 1946 W. Q. Kennedy claimed that the NE–SW Great Glen Fault, which has come close to turning northern Scotland into an island, showed at least 100 km of past horizontal movement.[4] For Wilson, who had spent his life mapping both geology and geophysics across the bare expanses of the Canadian Shield, these faults seemed at last to offer an insight into drifting continents. While mapping Newfoundland, Wilson had discovered that there was a great fault zone that could be traced as a major topographic and geological feature along the coast to the south-west as far as Boston.[5] On Wegener's reconstructed North Atlantic this fault matched with the Great Glen Fault of Scotland. The Canadians had finally proved their Caledonian forebears.

As a result of his position within the organisation of the International Geophysical Year and subsequently as president of the International Union of Geodesy and Geophysics, Tuzo Wilson had become very influential. In April 1963 he chose to write up his new inspiration titled, euphemistically, without fear of sensation, 'Hypothesis of the Earth's behaviour' in *Nature*,[6] and more directly as 'Continental drift' in *Scientific American*.[7] Within the positivist traditions of the latter magazine, the production of an article on a subject before its general scientific acceptance was extremely unusual, and provided powerful propaganda. The public was about to be informed that, like some audacious comet on an eccentric orbit, drift was, if not yet finally established, at least a glimmer in the eastern sky. Unwittingly, but as if to taunt Taylor's descent into obscurity, Wilson has chosen to name the fault presumed to pass through the straight-sided rift running between Ellesmere Island and the north-west coast of Greenland, not after the Indiana glacial geomorphologist who had first claimed the feature as demonstrating drift, but after his rival, Wegener.

In 1963, as a result of an invitation from Brian Harland, Sir Edward Bullard finally and unequivocally gave his full support to continental drift in a talk given to the Geological Society of London.[8] Bullard announced his intention 'to break away from the well-trodden circle of ideas' and discuss the new types of investigation that had grown up since the war. 'There are many phenomena, such as thunderstorms, ice ages, and the Earth's magnetic field that if they had not been observed, would probably not have been predicted by physical theory' Bullard remarked in defence of some mechanism, before discussing mantle convection and rock creep. In conclusion, drift is termed 'exceptionally fruitful'; if 'large programmes of investigation on land, and at sea . . . are pursued vigorously for ten or twenty years it is probable that general agreement will be reached'. The geological audience was less optimistic: one professor remarked that he 'regretted the lecturer's assertion that geophysical theories like continental drift were not obliged to consider the facts of structural geology.'

The article in *Scientific American* and the reveille for the geologists, although heralding a reawakening for the debate, were of no significance to the all-important process of scientific validation. The turning point for scientific discussion came at a two-day symposium organised by Blackett, Bullard and Runcorn at the Royal Society of London in March 1964.[9] All the old arguments that based themselves on geological evidence, on Palaeobotany and Palaeoclimatology, had now become reduced to one talk among twenty, placed at the beginning, in which it was admitted that 'purely geological evidence cannot prove or disprove drift'. Thereafter, the meeting concentrated on the new data. Bullard produced his long-awaited computer reconstruction of the continents around the North Atlantic. Jeffreys, on inspecting Carey's matched continents of the South Atlantic, had concluded: 'I simply deny that there is an agreement'.[10] Bullard is now employing the best-fit depth of 500 fathoms (914 m), liquidating Iceland for being too young, but turning Rockall and its surrounding shallows into fully fledged continental crust. Bullard achieved something that no human could have gained by laboriously tracing outlines over the globe; the maps had become blessed by the mystique of the computer (see p 149).

J. T. Wilson began the second day with the hammer-blow of argument striking hard and true: 'If the continents have moved, then they have drifted like rafts and formed the ocean floors in their wake. It is to this wake that we should look first.' After Wilson there came more indirect evidence of convection: the chemistry of volcanic magmas, radio-isotope age-dating to correlate reconstructed continents, tensional crustal drift continuing in Iceland; and so into the physics of convection currents in the Earth's mantle. The science was always there; the more geological elements of individual contributions had been analysed, age-dated with equipment that provided, however unreal, a sense of accuracy and certainty. Heezen still championed the expanding Earth; Menard and Vacquier are wary in their speculations. Suitably the most strident note of hostility was sounded by Worzel, Ewing's ambassador from Lamont.

American geologists and geophysicists attending this meeting were to claim on return that a balanced opinion had been impossible in this climate of intoxicated mobilism. Yet the enthusiasm that the British geophysicists now had for drift was still tempered by the sense that the theory lacked ultimate verification. The impact of the magnetic reversal story at this time can be gauged from Bullard's failure in his concluding remarks even to mention Vine and Matthews' paper. Sea-floor spreading was implicit in a number of the contributions, but only Vacquier introduced the new explanation for magnetic anomalies, claiming that 'this attractive mechanism' was not competent to explain all the details of the pattern found in the northeastern Pacific.

Serving up the plates

While the new information on moving continents was largely British, the fixist Americans, through their considerable oceanographic research programme, maintained overall control over the orthodoxies of the origins of the oceans. The tradition of more intellectual, less practical research attracted to Cambridge at the beginning of 1965 the leading theory builders: Wilson arriving in January, soon followed by Harry Hess on sabbatical leave from Princeton. A major marine geophysics cruise was underway and most of the researchers were at sea. Vine, however, had been left ashore to complete his Ph.D. thesis. In January, in the rambling airy rooms of Madingley Rise, the new science of the Earth finally found its intellectual hot-house.

On arrival Hess had offered Vine the most significant encouragement he had yet to receive on the magnetic anomaly and sea-floor spreading hypothesis. In contrast Wilson knew that Vine was working on mid-ocean ridges but had not read the neglected 1963 hypothesis. All three were intrigued by the problem of the great horizontal faults in the oceans; from the late 1950s, Heezen and other cartographers of the ocean floor had discovered that the mid-ocean ridges were commonly offset by up to several hundred kilometres along fracture zones. These fracture zones seemed to be oriented roughly transverse to the line of the ridge; all those other manifestations of the ridge from axial rift and magnetic anomaly to heat flow were likewise displaced. In 1963 the Lamont seismologist Lynn Sykes published the first results of the WWSSN network for the South Pacific which showed that the seismicity followed both the ridges and the fracture zones offsetting the ridges, and also that the fractures that extended into the oceans away from the ridge were free from seismicity.[11]

What happened to these fracture zones where they encountered the continents? Vacquier believed he could trace the Mendocino fracture zone into Nevada, and in common with Heezen and other researchers thought that the fracture zone must have once been active as a transcurrent fault along its length. Wilson went away on a two week sailing holiday of South Turkey and returned having solved not only the problem of the localisation of the active seismicity but had also extended the whole concept of magnetic anomalies and sea-floor spreading. In the climate of disregard for their theory, Matthews and Vine had never considered the implications of the idea: that the anomalies should be arranged symmetrically to either side of the spreading ridge. The second result was the more remarkable. Wilson had been playing with a paper sea floor. After dividing a single sheet of paper along a Z-shaped cut, it was possible to separate the two pieces through sideways 'strike-slip' movement along the central bar of the Z. The upper and lower bars were the spreading ridges. From paper cut-outs to the ocean floor; the fracture zones should only present seismicity in between the offset sections of ridges. Away from the spreading ridge the fracture zones are not seismically active because they are only fossil sites of movement. The active sections he named 'transform faults'. Together with Vine he began to consider the implications of such behaviour.

After the success with the ocean islands, Wilson's fascination with the giant fault zones had led him to undertake a major library research effort. As the San Andreas Fault was still active it became the most important focus of the study. Structural geologists had always been puzzled by faults of this kind, explaining their disappearance at either end through a process of 'horse-tailing', dissipation into a fan of smaller faults. While a student at Princeton, Wilson had heard Bucher lecture on the San Andreas Fault, claiming the pattern of earthquakes indicated that the fault continued offshore to the north-west for some distance beyond Cape Mendocino before suddenly and unaccountably coming to an end. In Runcorn's 1962 book *Continental drift* there was information presented about the active faults and earthquakes offshore from the Queen Charlotte Islands further up the Canadian coast. In an act of great symbolic convergence, one morning Hess had suggested that if a spreading ridge was required between these two sections of faulting then there should be some magnetic expression. Mason and Raff's map remained the only detailed study of a wide region of sea floor. Vine fetched the map and together they combined WWSSN earthquakes and magnetics. In the gap between the two sections of seismicity lay a region in which the magnetic anomalies were skewed but aligned around an ocean ridge first identified by Menard in 1955. Here the transform faults of the San Andreas and the Queen Charlotte Islands are more extended features than the intervening spreading ridge itself. Wilson wrote up his explanation for the pattern of transforms and spreading ridges[12] and in collaboration with Vine described the feature that had evaded all those thousands of scientists who had inspected the Mason and Raff map in the four years since it had been first published.[13] Around the ocean ridge, named by Vine and Wilson the Juan de Fuca ridge, the magnetic anomalies were symmetrical. There was a mirror in the midst of the ocean floor.

In early 1965 at Cambridge, the foundation had been established for a new understanding of the behaviour of the outer shell of the Earth. With the assistance of Vine and Hess, Wilson had coupled sea-floor spreading with seismicity and married the magnetic pattern of the ocean floor with the giant transcurrent faults on the margins of the continents. The distribution of seismicity that was apparent in the WWSSN maps indicated to Wilson that the majority of the outer shell of the Earth behaved with considerable rigidity and that movement was restricted to certain relatively narrow zones. These rigid blocks which moved relative to one another at the spreading ridges and transforms he termed 'plates'.

In their joint paper on the Juan de Fuca spreading ridge, published in October 1965, Vine and Wilson attempted to measure the spreading rate through correlating the magnetic anomalies and the latest timescale of magnetic reversals. Hess had by now modified his original picture of oceanic crust to allow a thick capping of basalts of high magnetic susceptibility to rest above his original low magnetic serpentinite. With the aid of a computer it proved possible to model the magnetic profile for a variety of spreading rates. The figure of 2 cm per year provided the most plausible agreement between the theory and the actual measurements.

The great conversion

Having finished his thesis, Vine was offered a research post at Lamont but preferred to accept an invitation from Hess to become an instructor at Princeton, commencing in September 1965. In November he was invited by Wilson to co-present the Juan de Fuca ridge paper to the American Geological Society meeting in Kansas City, Missouri. Brent Dalrymple was also there and on 4 November presented the most recent update on the magnetic pole reversal timescale,[14] in which he revealed that a new short-lived reversal appeared to have been discovered at only 0.9 million years. Vine met Dalrymple and sought to obtain more information on this reversal, for he had realised that with this new information, the anomaly pattern symmetrically arranged around the Juan de Fuca spreading ridge could be far more easily explained with a constant spreading rate than had been possible with the reversal scale that he and Wilson had already employed.

Back at Lamont, business was as usual. In 1963 James Heirtzler, the leader of the marine magnetics group, had convinced the US Navy to employ the sophisticated equipment aboard its own magnetic survey plane to fly a series of 58 lines across the Atlantic ridge immediately to the south-west of Iceland. The American base at Keflavik, near Reykavik, offered the electronic navigation and flight support. An area measuring 200 miles (320 km) along the ridge, and (200 km) across was completed and revealed a symmetrical pattern of anomalies, but the Lamont group chose to emphasise that the intensity of these anomalies decreased away from the ridge and that there seemed to be two classes of features present, and thus that the Vine and Matthews hypothesis was 'untenable, at least in its present form'.[15] In 1965

Heirtzler, and Xavier le Pichon, a visiting French marine geophysicist, wrote in collaboration with other Lamont researchers a total of three papers disparaging the possibility of mobility recorded in the sea floor. 'It is clear from this study' they had announced in one of these 'that most of the profiles do not follow the pattern assumed by Vine and Matthews'.[16] However, amidst their ranks there was a mobilist who, unlike Heezen, had not become diverted by the expanding Earth.

Neil Opdyke had studied Geology at Columbia before becoming Keith Runcorn's field assistant on the tour of America that was to provide Runcorn with the samples indicating that the magnetic polar wandering curve for North America had diverged from that of Europe. In 1955 Opdyke travelled to Cambridge to begin a Ph.D. and had followed his supervisor when Runcorn transferred to the University of Newcastle. From their earliest conversations, while driving out to the American West, Opdyke had been brain-washed with the English work on Palaeomagnetism. By the end of the 1950s, in company with all the rest of the English palaeomagnetists, Opdyke had become a convert to continental drift. In January 1964, on the advice of Ewing, who recognised the gap in palaeomagnetic studies in existence at Lamont, Opdyke obtained a research position under the supervision of Heirtzler. Having intended to continue his work on land samples, Opdyke was diverted into examining the palaeomagnetism of deep-sea cores. Towards the end of 1965 he had found evidence for a magnetic reversal at around 0.9 million years.

Walter Pitman III was a Columbia-trained graduate student under Heirtzler, fully sharing the Lamont fixist orthodoxy. From September to November 1965 he was engaged on a cruise on board the National Science Foundation vessel *Eltanin* around the Pacific–Antarctic ocean ridge, which had no intention of setting out to confirm or deny any hypothesis but simply, in the style of all Lamont cruises, to amass more information. The magnetic data were collected in analogue and digital form but not studied on board. The magnetics on Leg 19 achieved on a traverse across the East Pacific Rise at 51°S, immediately before Pitman had joined the ship, as well as Legs 20 and 21, were all to form part of his research project. On return Pitman processed the magnetic profiles obtained on Legs 20 and 21, removing the Earth's own magnetic field, and while assessing the significance of the results came across Vine and Wilson's paper on the Juan de Fuca ridge, newly published in *Science*. The Eltanin (Leg) 20 profile looked remarkably similar to a profile across Juan de Fuca. 'Ha ha' said one of the Lamont staff members 'I suppose you are going to prove Vine and Matthews are right'.

Pitman found himself introduced into an entirely unknown world: the papers of Hess, Dietz and Vine and Matthews. One month later the profiles from Leg 19 had also been processed, and Pitman, who had by this time gained a considerable induction into drift from Opdyke, pinned the new profiles up on Opdyke's door. 'When I came back'

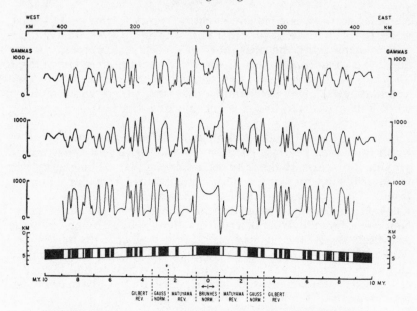

Fig. 3. The middle curve is the *Eltanin*-19 magnetic-anomaly profile; east is to the right. The upper anomaly profile is that of *Eltanin*-19 reversed; west is to the right. On the bottom is the model for the Pacific-Antarctic Ridge. The time scale (millions of years ago) is related to the distance scale by the spreading rate of 4.5 cm/yr. The previously known magnetic epochs since the Gilbert epoch are noted. The shaded areas are normally magnetized material; unshaded areas, reversely magnetized material. Above the model is the computed anomaly profile.

The Eltanin 19 magnetic anomaly profile (from Pitman & Heirtzler 1966, with the permission of *Science*, copyright 1966 by the AAAS)

remarked Pitman later 'the guy was just beside himself! He knew that we'd proved sea-floor spreading!'

Eltanin 19 was the first magnetic profile to preserve perfect bilateral symmetry across the ocean ridge. The two sides could be folded on top of each other. Pitman was already convinced; it took several weeks before Heirtzler could be brought round to accept sea-floor spreading. At first he had claimed that the profile was too perfect and could be explained only by some pattern of electrical currents in the upper mantle. When Heirtzler had been converted he had shown Eltanin 19 to Worzel. 'That knocks the sea-floor spreading nonsense into a cocked hat' claimed Worzel, before leaving the room; 'it's too perfect'. The Eltanin 19 profile fitted exactly the new magnetic pole reversal timescale; more significantly for a scientific argument, on the assumption of a constant spreading rate it *was* such a reversal timescale. Sykes was called in by Heirtzler to examine the profile and then to reassess the seismological evidence. Lamont was beginning to become transformed from the bottom upwards.

In February 1966 Vine came to visit Neil Opdyke, unaware of the turmoil that had overtaken Lamont. The walls of the office in which they met were bedecked with Eltanin profiles. Opdyke mentioned to Vine the new 0.9 million year reversal they had discovered in the deep-sea cores, and it was left to Vine to transmit the news then unknown to the Lamont palaeomagnetists, that Cox, Doell and Dalrymple had already discovered and named the event 'Jaramillo'. Vine was shown and was spellbound by Eltanin 19, subsequently requesting a copy, which was sent by Heirtzler with an accompanying letter remarking only that 'several of my students are using it in their theses'. As a result of some tender negotiations carried on through the year, it was agreed that Pitman and Heirtzler should publish the profile in a December issue of *Science*[17] two weeks before Vine presented his major 'Spreading of the ocean floor – new evidence',[18] the confirmation of all his earlier ideas.

Eltanin 19 was given its first airing to a small audience at the April meeting of the American Geophysical Union in Washington, DC. The magnetic reversal timescale investigators from Menlo Park were also at the meeting. Cox was shown the profile shortly before Heirtzler's lecture and suffered a 'truly extraordinary experience', recognising not only the whole of their timescale but also earlier reversals of which he had only the slightest intimation. Doell simply looked stunned and, summoning the full force of Californian articulacy, remarked 'It's so good it can't possibly be true, but it is.'[19]

The impact of Eltanin 19 at this April meeting was that of the appearance of stigmata to the infidel. The conversion to drift now began to spread as fast as Eltanin 19, the miracle of symmetrical anomalies, could be passed around. Its fame attracted a larger audience for meetings at the Geological Society of America in November 1966 in San Francisco. Cox, Doell and Dalrymple presented their perfected scale of pole reversals, and Vine his equivocally unequivocal 'Proof of ocean-floor spreading?' On the East Coast in November, Lamont researchers underwent a mass conversion at a meeting in New York sponsored by NASA at which all the data collected by Lamont on magnetic lineations and seismicity were combined. On the first day Ewing had remarked anxiously to Bullard 'You don't believe all this rubbish, do you, Teddy?'[20] At the end of the final session Bullard was to sum up in favour of drift with a second speaker against; but when the second speaker was unable to attend, no one volunteered to take his place.

Bullard spoke of 'how far we had gone from the traditional geological interest in the continents and their mountain systems' and recommended a return to these problems. Ewing was not to be finally convinced of the spreading sea floor until three years later. Among marine oceanographic institutions Lamont was the bunker of fixism. On a subsequent visit, Runcorn, the British geomagnetist who had advocated drift from before 1960, remarked 'I feel like a Christian visiting Rome after the conversion of Constantine'.[21]

In January 1967 nearly 70 abstracts on sea-floor spreading had been submitted to the Washington meeting of the American Geophysical

Union to be held in April. Four sessions in four days were devoted to the ramifications of the theories of the new dynamic ocean crust and the ideas had already spread into talks on seismology, tectonics, petrology and geochemistry. Even the largest auditorium proved too small for Harry Hess's 'History of sea-floor spreading'.

The geometry of the spheres

All the elements of a theory of the Earth were now available, only requiring a plan to become assembled. Geometry was about to reassert itself in the configuration of the globe; not the intersecting planes of the crystallographers that found their apotheosis in Elie de Beaumont's cobwebbed planet, but instead the ordered geometry of the great 18th-century mathematician Leonhard Euler. 'Euler calculated without apparent effort, as men breathe, or as eagles sustain themselves in the air' reflected the French mathematician D. F. J. Arago. While living and working in St Petersburg, amidst his hundreds of novel formulae and algorithms, Euler conceived, and published in 1776 at the age of 69, that any motion of a sphere over itself can be regarded as a rotation about some axis passing through the centre of the sphere. Where this axis meets the surface is the pole of rotation.

It was the memory of his lectures in undergraduate mechanics that led Bullard back to discover the theorem outlined in a textbook: Routh's *Rigid mechanics*. It was with this formula that Bullard, with his researchers Everett and Smith, had set about testing the fit of the continents to either side of the Atlantic. Their choice of the pole of rotation 'seems to have established itself'.[20]

Once Wilson in 1965 had shown that those regions of the crust free from seismic activity could be considered as rigid plates that moved relative to one another, the way was open for a treatment of the whole globe according to Euler's theorem. Yet Wilson himself did not follow through into the geometry: had he in his desire to stress the rigidity of the units forgotten about the globe? The validation of sea-floor spreading led to an immediate need to reassess all the implications of Wilson's 1965 paper on the transform faults. How did the motions of the plates fit together for the whole planet? This was no longer a time for the wise and powerful department directors to sustain a general interest in continental drift. After traversing a difficult mountain pass a whole new world had presented itself. It was a time for the young researchers, for those who could give up everything in order to work diligently, urgently, to win the race to be the first to show how these new principles made sense of the globe. Enormous quantities of data, already collected, had now to be sifted and reinterpreted. An information implosion followed on the Big Bang.

The combatants were new but the institutions, the milieu in which they worked, remained the same: Princeton Geology Department, the

The pattern of transform faults in the equatorial Atlantic as compared with arcs of circles centred on the pole of relative rotation between South America and Africa (at 62°N 36°W) (from Morgan 1968[25], with the permission of the American Geophysical Union, copyright AGU)

Department of Geodesy and Geophysics at Cambridge, Lamont-Doherty Observatory, Columbia University. Only Wilson had no inheritance of energetic researchers to explore the new science. To fulfil the task it was necessary to be a researcher at the very beginning of a scientific career. William Jason Morgan was well placed, working between 1964 and 1966 as a research associate on completion of his doctorate in Princeton Geology Department, having originally studied Physics. His first appointment as an assistant professor began exactly a year after Vine had arrived to take up his post as an instructor. Yet Morgan was in competition with another researcher whose academic life had also run one year behind and parallel to Vine's but for the two years before Vine had reached America. Dan McKenzie, born in 1942, had completed his Ph.D. at Cambridge in 1966, by which time he was already a King's College research fellow. The topic on which he had written the doctorate, 'The shape of the Earth', had prepared him to think of the physics of the whole planet, and in particular of its surface as controlled by hidden motions. Both Morgan and McKenzie, working and conversing with the creators of sea-floor spreading, had a head-start on all rivals. Close behind there were the Lamont converts who, after Eltanin 19, worked like zealots, men possessed, to make up for the time lost supporting now discredited scientific orthodoxies.

The race for priority in publication was won by McKenzie. In collaboration with R. L. Parker at the Institute of Geophysics and Planetary Physics at San Diego where McKenzie was a visiting researcher, the paper was published one year after Vine's celebration of the truth of sea-floor spreading, and most significantly two days before the new year, therefore forever after making it McKenzie and Parker (1967).[22] The days between the manuscript's arrival at *Nature* on 14 November and its eventual publication must have been filled with nervous anticipation of being pre-empted for, unlike the confirmation of sea-floor spreading with Eltanin 19, the global theory required no more information than was already available to other researchers.

McKenzie and Parker restate Euler and then recognise the significance of transform faults: transform faults must be arcs of circles drawn around the pole of relative rotation of the plates. Where three plates meet the possible plate motions become constrained and can be represented as a vector triangle of velocities. With the knowledge of only three parameters of the triangle including at least one side (or velocity), it is possible to solve all the other unknowns. The geometry of high-school, Euclid.

In 1965 Wilson had shown how ocean-floor magnetics could be integrated with the concentrations of seismic activity. McKenzie and Parker did not have any direct access to data on the ocean-floor magnetics, so instead they chose to study the nature of the fault movements of plate-boundary earthquakes as detected by the WWSSN network. Their research had focused on the largest of the Earth's plates: the Pacific to the north of the East Pacific Rise, occupying about one-quarter of the planet's surface. The fault movements were entirely consistent with what they term 'the paving stone theory' for rigid plates.

Although they did not mention it these results were not new. A decade before a Canadian-trained geophysicist, Adrian Scheidegger, a protégé of Wilson, without the benefit of WWSSN, found from his study of the direction of fault movements extracted from the instrumental record of 30 years of earthquakes, that 'the *whole* Northwest Pacific Ocean is moving *en bloc* with regard to its margin'.[23] 'This' wrote Scheidegger 'is a rather startling result'. 'This' we now know is one of the simplest demonstrations of plate tectonics. Scheidegger did not see the global significance of his observations nor the Eulerian geometry. He was, he admitted, despite 'great sympathy' with drift, 'after much thought and disappointment', a geodynamical 'agnostic'. He should have read a paper published in 1957 by an English geologist, Albert Quennell, who indicated an Eulerian pole of rotation to explain the opening of the Red Sea.[24] The pole of rotation of the Pacific plate relative to North America, as located by McKenzie and Parker, lay in Canada, immediately to the south of the Hudson Bay.

Morgan's paper arrived less than three months later[25] but already contained far more detail. Morgan revealed several of the relative motions of what he preferred to call the 'rigid blocks' as well as a map showing the 20 'such units' that make up the Earth's surface and a number of their poles of relative rotation and speeds of relative movement. The blocks, he finds, must be thicker than the ocean crust. The rigid units, perhaps 100 km thick (Barrell's lithosphere) slide over a weak asthenosphere, now claimed as the unseen lubricant that allowed the blocks to glide across the globe. As if to distance himself from Wilson, Morgan prefers 'block' to 'plate'; a term too banal to become widely emulated.

In writing his paper, Morgan had had access to a publication of the

Lamont seismologist Lynn Sykes, which also appeared in 1967,[26] that confirmed from the evidence of both the location and the motions of the earthquakes of the mid-Atlantic all Wilson's 1965 theory of transform faults. More significantly, Seismology had the opportunity to investigate the greater mysteries of those locations where, if the theory was correct, the ocean lithosphere is being consumed. Since 1964 the Lamont seismologists had concentrated their attention not on the shallow earthquakes employed by McKenzie and Parker but the configuration of deep earthquakes located far beneath the Tonga islands.

The disappearing oceans

Deep earthquakes were first recognised at the beginning of the 1920s and found to be concentrated beneath the arcs of volcanoes fringing the oceans. In 1934 the Japanese seismologist Wadati had shown that the earthquakes formed an inclined plane dipping away from the ocean[27] and subsequently Vening-Meinesz had used the evidence of such deep underthrusting to extend his theories about local and episodic convection. The large majority of these earthquakes lay around the margins of the Pacific.

In 1954 the Californian seismologist Hugo Benioff published a cross section of the seismicity beneath the Kamchatka Peninsula and the Kurile Islands that curve towards Northern Japan.[28] The intersection of the line of earthquakes with the surface came at the ocean trench, thus providing another important confirmation of the interrelations between seismicity and the most accessible measure of the oceans: the topography. The zone of earthquakes gained a number of different identities: in America, with no attention to anything other than national claims for priority, it became known as the 'Benioff Zone', a name that through *force majeure* has been able to overtake the most legitimate Japanese 'Wadati Zone', and appears to be fast displacing the Soviet 'Zavaritskyi Zone'.

Deep earthquakes extending down to 700 km had always seemed extremely unlikely, as the hot rocks at depth were expected to deform gradually, rather than through sudden seismic outbursts. The Tonga survey operated by Lamont had not only shown that the Benioff Zone of earthquakes was restricted to a band only 25 km thick but also confirmed a discovery first made by the Japanese seismologist Katsumata that high frequencies of seismic vibration travelled along the plane of the zone as if it was hard and cold. By late 1966, the Lamont seismologists encounter with Eltanin 19 had encouraged them to think in terms of sea-floor spreading. The high seismic frequencies also travelled with great efficiency along the ocean crust itself. What if the Benioff Zone were the site of ocean crust descending down into the mantle?

With all the attention concentrated on the mid-ocean ridges, the

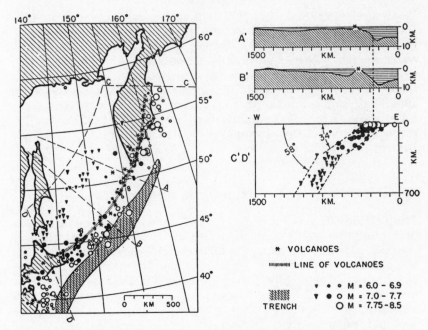

Benioff's 1954 plan and cross section of the seismicity of the Kamchatka–Kurile arc
(with the permission of the Geological Society of America)

trenches and island arcs had suffered considerable neglect. It was now for the first time possible to re-evaluate the successes and deficiencies of Carey's expanding Earth. All his separating sections of continents opened as either rifts or a set of jaws, and while the blocks that made up the jaws rotated, they did so about an axis lying at the edge of the opening. Thus his model was only partial; he had never recognised the general rule or Euler's theorem. Heezen had become converted to this expanding Earth because he lived in a world only half formed. In common with Carey he could perceive only where new ocean crust was created and not where it was taken away. It is a philosophical problem of identities. Imagine a child who had become familiar with the notion of birth but not of death. Creation, for the innocent child, is evidenced by people everywhere; destruction by what – by tomb-stones? For the whole Earth theorist while creation could be evidenced by the pattern of sea-floor magnetic anomalies, destruction made itself known more indirectly: Mount Fuji, Krakatoa, Mount Shasta, Popocatapetl, had to be seen as memorials to the death of the ocean floor beneath them.

The first programme of Marine Geophysics organised by Vening-Meinesz, had concentrated on the trenches and their large deficiencies in gravity; deficiencies that he claimed could only be sustained through downward movement. Ewing had later declared this evidence to indicate that trenches were tensional phenomena. The earthquakes of the Benioff Zone, the volcanoes of the island arcs and of the Andes; these gave evidence of some common process, but of what? When Hess had first proposed his spreading sea floor and underlying convection, Ewing had checked the profiles of the oceanic trenches to see if they were choked full of sediments stripped off the plunging ocean floor, and having found them relatively clean had concluded that sea-floor spreading was bunk. However, in the 1950s two guyots had been discovered partly foundered within separate deep-ocean trenches,[29, 30] evidence that was to become most powerful in the attempt to demonstrate underthrusting. At the end of the paper describing the earthquakes beneath Tonga, Isacks, Sykes and Oliver had stated with surprising self-confidence 'the seafloor is being pushed, or pulled, down into the mantle to form the zone'.[31]

The desire to make the new theory compatible with geological uniformitarian principles and create, like Hutton, a world involved in an exactly balanced cycle of creation and destruction, allowed the Benioff Zone to be considered as the site of the destruction of ocean crust without a great deal of dispute, except, of course, from supporters of the expanding Earth. For the greatest argument in favour of ocean destruction is identical to that employed by the (rational!) child naïve of death: 'No people (ocean crust) in my experience have ever been found to be older than, say, 100 years (200 million years), yet the world is, I believe, far older than this. Therefore either I, the child, am part of the chosen second or third generation to have appeared on this planet or else people (ocean crust) somehow cease to be'. Warren Carey has metaphorically denied death. He has also denied uniformitarian principles; which may seem less outrageous. Yet Hutton had formulated the greatest contribution to Philosophy to have emerged from Geology: in the manner of the Buddhists, man was to become an incident in the universe, a drop of water in a river, against the Judaeo-Christian vision of a world in orbit around man. The perceptions confirm the Judaeo-Christian view, but the study of the Earth has always had to fight against the evidence of the senses. The expanding Earth, through its need to have a process of expansion accelerating through time up to the point when man could be here to observe it, has again wrapped the world around man. Like Geology itself, the theory is egogeocentric. The continuing survival of many advocates of an expanding Earth is that most unlikely of all fossils; not the fossil of the original culture which survives within Geology, but of part of the emerging culture, like some intermediate state in a complex chemical reaction, frozen early on during the process of transformation.

In June 1968 the French marine geophysicist Xavier Le Pichon, still a visiting researcher at Lamont, achieved atonement for his vigorous denial of drift published only two years before, by producing 'Sea-floor spreading and continental drift'.[32] From a study of magnetic anomalies and orientations of the transform faults, Le Pichon had gained the poles of relative rotation and spreading rates for the six most important 'blocks'. (Like Morgan, Le Pichon, too, is shy about 'plate'.) Through a comparison of the extended pattern of magnetic reversals revealed both in deep-sea cores and in profiles across the oceans, he extrapolated spreading rates back through the Tertiary era, finding the configuration of the continents imbedded in these rigid blocks for the past 200 million years.

In September 1968, the Lamont seismologists Isacks, Oliver and Sykes presented their own contribution: 'Seismology and the new global tectonics',[33] in which for the first time the zones of convergence were given an equal status to the zones of spreading. The new global tectonics that they proclaim incorporates seismological information with other branches of Geophysics and Geology. Yet the paper is strangely gawky, and still the climate of Lamont uncertainty persists. Plates are scarcely mentioned: 'Implicit in the new global tectonics is a new attitude toward mobility in the earth's strata' they state confidently, but then the final sentence: 'Even if it is destined for the discard at some time in the future, the new global tectonics is certain to have a healthy, stimulating and unifying effect.' The influence of Lamont's sceptical director still permeated the output of his researchers.

'Stimulating', 'unifying', 'healthy'? In March 1969 McKenzie sent to the Royal Astronomical Society 'Speculations on the consequences and causes of plate motions'[34] in which he assessed all the evidence available for consumption of ocean floor, and went on to discuss the role of sinking oceanic lithosphere in promoting convection.[32] In October 1969 he had teamed up with Morgan to write a detailed and definitive paper on the surface geometry of plate triple junctions, in which through the influence of Morgan his earlier 'plate theory' has become 'plate tectonics'.[35]

The final touches were put to the theory in 1970. While the Americans had initially been reluctant to use the word 'plate', Wilson and the Cambridge axis of Vine and McKenzie had promoted the term into currency. The Benioff Zone became 'the subduction zone'; subduction being an archaic term for subtraction, but unlike subtraction (literally 'to draw under') subduction, 'to lead under', was innocent of cause and effect, push and pull.

Field's original intention to allow Geology to conquer the oceans had become reversed; the race of intelligent amphibians, the marine geophysicists, were now about to crawl onto and conquer the land through a rapid phase of evolution and adaptation. Plate tectontics was finally going to offer a solution to the riddle of the mountain ranges which had kept geologists engaged in a tail-chasing debate over geosynclines for 100 years.

In 1966 in an article published in *Nature* Tuzo Wilson had posed a rhetorical question in the title 'Did the Atlantic close and then re-open?'[36] For around the North Atlantic there lay fossil clues that were as puzzling but also as simple as any that could be imagined. During the Ordovician and Silurian geological periods, amidst the shallow marine sediments, there were two distinct groups of fossils, the products of two different faunal populations. Those faunal populations of North America were dissimilar to those from Western Europe but with some remarkable exceptions. For 'American' fossils are found in Scotland, north-west Ireland and coastal Norway, while 'European' fossils turn up in southeastern Newfoundland and along the southern coastal rim of New England. An ocean, Wilson claimed, an older Atlantic, had once separated Europe from America and, having closed, had re-opened along a line offset from its original suture, thus exchanging bits of geography: American Scotland for European Rhode Island: not even Snider–Pellegrini could have conceived of a myth so strange. To claim that oceans had once traversed some of the most familiar terrains, making England and New England kingdoms torn and rehealed was extraordinary indeed. To claim that oceans were the keys to understanding the greatest problems of Geology was outrageous: Lyell's race of intelligent amphibians had at last proved themselves superior to geologists.

In 1970, this story of united continents, was substantiated by a trans-Atlantic pair, John F. Dewey of Cambridge, England, and John M. Bird of State University, New York, who combined to provide a detailed reconstruction of the Appalachians.[37] Dewey had already begun the story the previous year.[38] The type geosynclines that had first been identified from that eroded mountain range, were no more than sediment that had built up on the margins of an ocean. The whole history of the Appalachians could be unravelled according to how it fitted into the pattern of past oceans and movements of the plates; of episodes of the loss of ocean floor down subduction zones and finally of continents colliding. A second collision with Africa had forced up the mountains whose ruins now run through Virginia and Pennsylvania. The loss of the ocean floor brought previously un-related sections of continental crust into contiguity. As plates could move only through the creation and destruction of oceans the new theory had less to offer this final stage, the Himalayan–Alpine stage of the orgy of mountain building, when in place of the simple consumption of ocean floor the continents must find new strategies to resolve their conflicts.

After unravelling the Appalachians it only remained to find this same kind of detail within other mountain ranges. Dewey and Bird in their 1970 'Mountain belts and the new global tectonics'[37] achieve this universality of explanation. There would be many problems remaining, problems where past horizontal movements brought into juxta-position sections of continent that could not be located in their original positions because only the loss of ocean crust left record of its passing. Yet so many large-scale geological problems could now become solved through the agency of the appearing and disappearing oceans. Suess had denied the existence of geosynclines because he could see no modern examples of huge piles of sediments developing in a continental interior. Now the problem could be solved through redefining its fundamental frame of reference: such continental

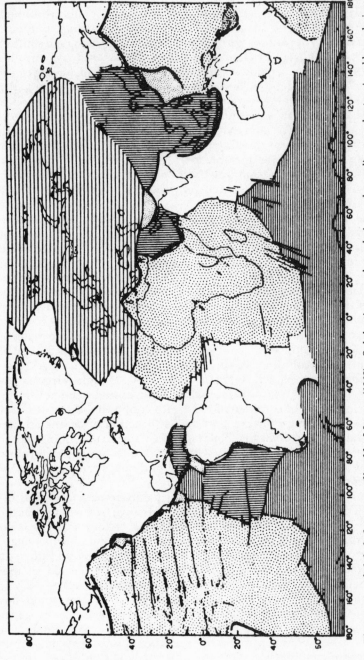

The pattern of global plates as outlined by Morgan (1968)   Subduction zone boundaries are heavy lines, and extensional boundaries are thin lines (with permission of the American Geophysical Union, copyright AGU)

interiors had been formed by the collision of continents after the loss of an intervening ocean.

Not only had writings on the plates begun to proliferate, but all the large-scale information previously collected by geologists was suddenly ripe for reinterpretation. By the end of 1970, as the language had consolidated, so workers in many fields of Geophysics and Geology had begun gingerly to employ the new terminology and to think according to a brand new scale: the scale of the globe. A whole-Earth theory had eventually arrived that was becoming ratified in all branches of the science. An explosion of new research began. Behind the island arcs of the Western Pacific, tensional marginal basins were discovered, floored with ocean crust that showed, not as Carey had assumed that the Earth was rapidly expanding, but that subduction can pull the crust apart to either side of the island arc.[39] The most fertile areas concerned the reinvestigation of geological problems of the continents; such as the evidence of strange rocks reminiscent of ocean crust, squeezed up onto the land.[40] Perhaps the vanishing, lost ocean floor did not all disappear into the Mantle?

Thus continental drift was seen to be true; the continents were not like ice-floes, but like logs imbedded in the ice-floes that were the plates. Continental drift had always been impossible to prove in its own right, because it was no more than the effect of greater hidden processes, the tip of the iceberg. Cause was hidden in the mantle. If all the plates were jostling over the Earth's surface, then was there no fixed reference frame? Morgan was to propose in 1971 that the age relations of the Hawaiian islands, first detailed by Wilson, reflected but one of many underlying 'mantle plumes' or hot-spots that might indeed provide some more permanent fixed configuration of volcanic scars as the plates glided past.[41]

As plate tectonics represented a great victory for the uniformitarians over the catastrophist geologists, who after Stille and Kober had sought to find world-wide rhythmic episodes of cataclysmic mountain building, so it became pertinent to ask about the limits of this uniformitarian behaviour. Did the plates ever stop moving and start again? Could there be intercommunication to provide world-wide tectonic events? What had preceded plate tectonics when the Earth was hot, young and the continents pliable? Having failed to conquer the Earth, catastrophism went back to its traditional source of inspiration in the stars. Catastrophes were once again to find their origin in the extra-terrestrial: in the collision of the Earth with comets and asteroids. In 1964 Dietz had identified that 2000 million years ago an asteroid had hit Sudbury, Ontario, carving out a crater 100 km across and 25 km deep and providing, in the subsequent magma-infilling, all that region's rich nickel deposits.[42] The battles were to shift ground; battles that seemed as primitive as differences in the human soul; that had once in the 14th century separated frightened apocalyptics from the merchants and peasantry who had faith in continuity.

The most steadfast demonstration of sea-floor spreading was to emerge from the deep-sea drilling project of the Joint Oceanographic Institutes for Deep Earth Sampling project (JOIDES), funded by NSF, which began in August 1968. The big three, Lamont, Woods Hole and Scripps, had finally allowed a young pretender, Miami, to join their club and so to initiate the programme that was the positive product of the Mohole project (see Chapter 8). Within a few years other institutes in America, Europe and the USSR had joined JOIDES, but meanwhile the marine geologists had come to realise the ultimate indignity of a great scientific breakthrough, that they were merely confirming what others had predicted: the radio-isotope dating of the basalts located beneath the sediments was proving at great expense what the marine geomagnetists already knew.

8

The age of the Earth

A revolution in the Earth Sciences!

The rites of passage that had greeted the birth of Geology were to be heard again, but now to celebrate a new arrival. The closer one gets to the birth the louder and more exuberant becomes the welcome; enthusiasm overtakes rationalism; the great new wave obscures all but the most jaded view of what is past.

The first book to be written in the light of the new world also celebrated the foundation of a progressive experiment in education; the Open University, created by a British Labour Government to provide undergraduate courses by television and correspondence. *Understanding the Earth* was to be the first-year textbook for the new Department of Earth Sciences. The book bares its soul in the preface, beginning 'during the last decade, there has been a revolution in the Earth sciences'. In the first four paragraphs 'Earth sciences' is repeated seven times; Earth scientists another three. The antecedent Geology is not mentioned except by implication in a justification for why the Open University found it necessary to commission an entirely new text: 'we needed a book which would support our course in emphasizing that the Earth sciences are alive and vital; that present-day Earth scientists are very far removed from the all too prevalent mental picture of elderly, bewhiskered gentlemen diligently brushing the dust from museum collections of rocks, minerals and fossils.' These by imputation are geologists. Earth scientists are young, clean-shaven, opportunist, ambitious, impetuous; the dust would have to be left to settle. Later there is a reference to 'geological data, painstakingly accumulated by land-bound geologists'. The adverb could have been derived from meticulous but instead it is 'painstaking'; the suffering was self-wrought, unnecessary.

How rapidly this transformation overcame the science and how closely it is possible to locate the historic watershed can be found from an important textbook published earlier in 1970 with a preface written in November 1969, exactly one year before that of *Understanding the*

179

Earth. In emulation of Jeffreys, and without the promise of enlightenment offered by the title of the Open University Text, this work was
simply *The Earth*; produced by a group of geologists from Berkeley.

In a book of 750 pages, the new theories merit only 20, lost in the
midst of Chapter 13 on the 'Structural evolution of continents and
oceans', in which continental drift and plate tectonics are described as
provisional theories in danger of deflation should a few further
observations undermine them. *The Earth* and *Understanding the Earth*
are separated by 12 months, ten thousand kilometres, and a total
conceptual transformation. In *The Earth* there is no celebration of a
new science; instead the preface announces that the work is intended
to 'introduce the reader to physical geology'; a subject that one year
later according to the circumspection of *Understanding the Earth* had
little further relevance.

The preface to *Understanding the Earth* began with an affirmation that
was, above all others, to become the slogan of the transformation to
plate tectonics: 'A Revolution in the Earth Sciences'. This phrase was
first, and most appropriately coined by Wilson. While President of the
International Union of Geodesy and Geophysics at the 1963 General
Assembly in Berkeley he delivered a rousing incitement to overthrow
the old order: 'I believe that earth science is ripe for such a revolution;
that in a lesser way its present situation is like that of astronomers
before the ideas of Copernicus and Galileo were accepted.'[1] Four years
later, in a lecture delivered in Canada, he was to announce that: 'the
revolution in the Earth Sciences' had now taken place.[2]

All that had *happened* during the 1960s, from the renaissance of drift,
allied with sea-floor spreading, the plates and their new geometry,
was to be so described. In 1973 A. Hallam wrote a book with this
statement as its title. 'A revolution in the Earth Sciences' was a
portmanteau phrase so widely repeated that it never suffered investigation. If analysis is demanded, then all attention is directed to the
'revolution'. Yet one of the first acts of any revolution is to create a new
order through providing new names and new institutions. It is far
easier to change St Petersburg into Leningrad than to reconstruct the
fabric of an imperial city into a workers' communard. Let us look again
at the 'revolution in the Earth Sciences'. The rhetoric of revolution we
accept; but what of the Earth Sciences?

The preface to *Understanding the Earth*, which fails to mention
Geology but hammers home Earth Sciences, is a clue. If the revolution
was *in* the Earth Sciences, what was happening in the Earth Sciences
before the revolution? The *Oxford English Dictionary* does not list
'Earth Sciences'; an investigation of other dictionaries reveals that
'Earth Sciences' has no currency until the mid 1960s. Academic
journals, always responsive to innovations of thought, incorporated
'Earth Sciences' in their titles with 'Earth and Planetary Science
Letters' and 'Earth and Planetary Science Reviews' both initiated in
1966. The *Encyclopaedia Britannica* has no entry in 1970 but by 1974
contains a section on 'the Earth Sciences' that runs for many pages.

As a response to something that was happening to Geology in the 1960s, North American Departments, Schools and Colleges of Geology began to change their names. Probably the first of all these name changes came at Wilson's own department at Erindale College in Toronto; in the USA the first School of Earth Sciences appeared at Stanford just before 1965. By 1970 there were four such departments, by 1975 seven; by 1980 at least ten. Earth was sometimes conjoined with 'Space' as at UCLA, or with Planetary, as at Johns Hopkins; or even the strange tautological 'Earth and Mineral Sciences' at Penn. State, reflecting no doubt some stratagem to placate obstructive mineralogists. Many other departments and schools also changed their names, not to Earth Sciences, but to Geosciences, or Geological Sciences, or Earth Resources. Such name-changing provides a powerful demonstration of a new purpose and direction to the science.

Despite the silence of the dictionaries and encyclopaedias it is possible to trace the term 'Earth Sciences' back much further than the 1960s. The first man to have given it currency seems to have been, most suitably, Chamberlin, who wished to create a new Geology, superior to all other sciences, a science as grand and imperial as his own ambitions for his Chicago Department; a science that was to embrace Cosmology, Cosmogony and Astronomy. Chamberlin's enlightened and grandiose theories of education and science contain remnants of the philosophy that predates Geology, surviving in the mid-West plains from the German geognosists of the 18th century. Thus the prehistory of Geology sustained itself most tenuously into the science that was to come after.

Wegener uses the term 'Earth Sciences' in the introduction to the 4th edition of his *The origin of the continents and oceans*. While Chamberlin employed it to invoke a grander empire based around Geology, Wegener offers it as an alternative science that can hope to indicate the parochialism of Geology in its demands both to judge drift and to ignore all geophysical information. 'Scientists still do not understand' he wrote 'sufficiently that all the earth sciences must contribute evidence towards unveiling the state of our planet in earlier times and that the truth of the matter can only be obtained by combining all the evidence.' Thus it is Wegener who is a prophet of the new science to come.

After Chamberlin there was no geologist large enough in ambition to sustain any science so grand. Thus Arthur Holmes, as blessed by Jeffreys, was a geologist apart, and yet still fundamentally at heart a geologist. His great textbook, begun while on wartime fire-service in Durham, was titled in homage to the founding father of 19th-century Geology, Charles Lyell: *Principles of Physical Geology*. His attempts to save his beloved Geology through extending its sphere of activity were deeply radical. Like Chamberlin he viewed Geology as the most synthetic of all sciences, but in place of some ideal Earth Science he constructs a diagram (at the beginning of *Principles of Physical Geology*) in which he shows GEOLOGY, subtitled 'the Study of the

Earth' as a bureaucratic umbrella with the lesser subdivisions of Geophysics and Geochemistry on the margins. This is a subtle subversion; as a geologist he believed that the science of Geology should adapt by becoming more global in conception. His vision of the greater Geology is little different from that of modern Earth Sciences in scope but possesses a traditional intellectual hierarchy, with the old Geology remaining at the centre. Holmes's attempt to redefine Geology inevitably failed. A word, like a culture, cannot be wrought into a new shape. New words are needed to affirm new cultures, new intellectual spheres, simply to prevent the old ones with their weight of history, reclaiming and overwhelming the new.

In 1953 an ambitious French geologist Marcel Roubault renamed the provincial Annales de l'Ecole Nationale Supérièure de Géologie Appliqué et de Prospection Minière de l'Université de Nancy with the heady title 'Sciences de la Terre'. He wrote of his dream of liberation from 'un corset d'une réglementation étroite'.[3] However the successful attempt to name, and thereby create, an Earth Science came in the US. Its origins can be found in three separate developments all of which began in 1957 and none of which have any direct bearing on the discoveries of plate tectonics.

The creation of the Earth Sciences

The 1957 International Geophysical Year was born out of studies of the weather and of the Arctic. In 1851 the disabled American naval officer Matthew Fontaine Maury, initiator of both the US Naval Observatory and the Navy's Hydrographic Office, obtained permission to approach foreign embassies in Washington with a scheme to collect weather information at sea. In 1853 at an International Conference held in Brussels, the 10 most important seafaring nations agreed that their warships would collect weather and oceanographic data, on standard forms, to be pooled for the good of mankind. Even if the ships were seized in battle, these records were to be preserved. Maury was triumphant: 'Rarely before has there been such a sublime spectacle: all nations agreeing to unite and co-operate in carrying out one system of philosophical research. Though they may be enemies in all else, here they are to be friends.'[4] He next turned his attention to the more jealously guarded land frontiers; but was to learn a much-repeated lesson, that internationalism in science can continue until there is the nationalism-affirming state of war; in this case the war in the Crimea (1853–5).

In 1872 at a meeting in Leipzig, in celebration of the foundation of the new German state, the weather scientists gathered together to discuss the formation of an international society, the International Meteorological Organisation, that was constituted in 1878. Even before the first convocation of the organisation in Rome, an Austrian naval lieutenant, Karl Weyprecht, recently returned from explorations in the Arctic, had made proposals for the establishment of temporary weather stations, around the Arctic and Antarctic, to obtain synchronised observations. In the spring of 1877 the research programme, which was also to include magnetic observations, was fully detailed. Following the familiar pattern, internationalism was then interrupted by the Russian declaration of war on Turkey.

When the Rome meeting did eventually take place, these plans had already received the approval of Chancellor Bismarck, and an International Polar Conference was convened in Hamburg in 1880 with a second meeting at Berne. The First Polar Year ran from August 1882 to August 1883, involving independent scientific parties at 14 polar stations, all but 2 around the Arctic. Although it was international with 10 countries conjoining in research, this was to be a thoroughly Germanic research initiative. The science of Meteorology was on the ascendent at the same time as the German nation, and offered a scientific alibi for exploration and imperial dreams. As long as the German nation endured, Meteorology was to enjoy special support.

Even the disruption of the First World War failed to sever this patronage. The Second Polar Year was first proposed in November 1927 by Johannes Georgi, a German meteorologist and polar explorer, at a meeting of the hydrographic survey in Hamburg, where Wegener had formerly been director. The global links begin to be forged. Georgi had been the inspiration for Wegener's final expedition, was the last to see Wegener alive, and was later to write a short biography of the meteorologist.

Existing still in the umbrage of post-war scientific isolation, the Germans sponsored a conference in the palace of the Danish Parliament in Christianborg in 1928. Already the plans were far more ambitious than for the first Polar Year. The airship Graf Zeppelin, recently returned from a flight around the world, was to supply Arctic bases. The Russians endorsed the proposals, and a commission for the Polar Year was arranged to be held in Leningrad in 1930. Three research topics were identified: weather, magnetism and auroras. In spite of the Depression that undercut much of its funding, the Second Polar Year went ahead from August 1932 to August 1933. In the following years the Hydrographic Survey at Hamburg was given the job of compiling the year's daily weather charts, and had reached June 1933 when Hitler invaded Poland.

The International Geophysical Year, in effect the Third Polar Year, was first proposed at the natural divide of the 20th century, in 1950. On the evening of 5 April the American James A. Van Allen, a 35-year-old expert on rocketry who had been using some of the 69 captured German V2s to do research on virgin cosmic rays, organised a soirée of geophysicists to meet the renowned English geophysicist and Oxford Sedleian Professor of Natural Philosophy, Sydney Chapman. Chapman had been the first to construct a 'model' for the behaviour of the atmosphere and the enveloping space, that could begin to account for all the Earth's surrounding electrical and magnetic properties. He had also been the first Englishman to become enthusiastic about continental drift. Along with another one of those present, an atmospheric geophysicist, Lloyd V. Berkner, Chapman had been involved in the Second Polar Year. The conversation turned to the great potential offered by the new technical developments in Geophysics: the study of the oceans, atmosphere, electrical belts and internal magnetism; studies of the whole-Earth. Berkner proposed that the time was ripe for a new Polar Year.[5] (It was Berkner who in 1959 was to recommend the organisation of a world-wide network of seismograph stations.) Within a few months, and through a variety of International Commissions, the idea began to catch fire.

To coincide with an anticipated peak of sunspot activity, the 18 month period from 1 July 1957 to the end of 1958 was chosen for the study. In October 1952 at Chapman's suggestion the Third Polar Year was renamed 'The International Geophysical Year'. The Polar Years had been international almost by default, simply recognising every nation's inability to bring nationalism into the uninhabitable polar regions. The Geophysical Year was now to transcend political divisions, to be a science for all nations.

The success of the international co-operation in the lead up to the IGY may in part have contributed to a thawing in the Cold War, for after some initial reluctance the Soviet Union became one of the most enthusiastic and dedicated supporters of the scientific programme. The leader of the Soviet delegation to the IGY organising committees was Vladimir V. Beloussov, the Soviet grandmaster in the field of Geotectonics. While Wilson was president of the International Union of Geodesy and Geophysics, one of the principal scientific organisations controlling the IGY, Beloussov was the vice president. This conjunction was to have considerable implications for the future form of the plate tectonic debate.

The IGY was an immense success, flushed by the economic boom of the 1950s and the great wash of money passing into scientific research in the aftermath of the war. In a book inaugurating the IGY Tuzo Wilson had contributed a chapter that supported Earth contraction as the origin of the Earth's major crustal features, but is already pondering the interrelation between Geology and Geophysics: 'It will be apparent that new discoveries from geophysics are demanding a reconsideration of much geological dogma handed down from the last century'.[6] After travelling to all continents of the globe and visiting many of the research stations, his vision has become transformed. In 1961 he listed one among three of the year's great achievements (the other two concerning international co-operation): 'the transformation of earth science into planetary science'.[7] That in 1961 an earth science is still of the little earth and not already a planetary science indicates the global significance of this realisation and also its novelty. For Wilson it was the IGY that turned the little earth into the big Earth. By 1961 Wilson is writing favourably about deep Earth convection currents; he has passed from being a contractionist to a mobilist-agnostic about whole-Earth theories.

Almost before it had begun, any public excitement about the IGY became overwhelmed by the technological achievements of just one of the programmes listed as being part of the research effort. 'Except for the Japanese attack on Pearl Harbour, probably no event has taken the American people quite so much by surprise'.[4] The Soviet space effort, although announced, had been completely ignored in the West. On 30 September 1957 the co-ordinating conference for rocket and satellite plans for the IGY met for a week in Washington. On the final day of that conference the Soviets sent Sputnik into orbit. The much-publicised American Navy's Vanguard effort to launch a satellite had yet to be tested when on 3 November the Soviets announced that a second, much heavier, satellite was now circling the Earth.

The American press went in search of scapegoats for their nation's technological humiliation. On 6 December in a blaze of publicity the Navy ran the countdown for the untested Vanguard rocket at Cape Canaveral. On reaching 'First ignition!' it promptly exploded in full view of the television and press of the world. On 5 February 1958 a second Vanguard flew for 60 seconds before its control system sent the vehicle into a swerve so severe that the rocket fell to bits.

The Vanguard débâcle, and the loss of the first round of the space-race, forced the American government to support a massive space effort in order to attempt to overtake the Soviets. The civilian space agency NASA was formed in 1958 to take the programme away from the military.

In the year of extended international collaboration that followed on the IGY, the USSR launched three Lunik Moon probes. The first flew past the Moon, the second made a direct hit and the third flew round the back of the Moon, returning into Earth orbit with the first photographs of the Moon's 'dark side', which were transmitted to the ground via a television scanner. The operation of the Lunik missions within the extended IGY ensured that the Moon, like Antarctica, was to be located within a purely scientific exploration initiative. The response was of course a new American Moon initiative, launched by President Kennedy. To substantiate this endeavour as being more than the simple desire to avoid national humiliation in the space-race required that a new science was prepared to study the scientific data that were to be collected at such great expense. Central to this new research effort was to be the investigation of Moon rocks with a thoroughness that had never been employed for stones from the Earth. Although the first such rocks were not returned until the very end of the 1960s the prominence given to the journeys into space over the decade following the end of the IGY was to ensure that the views from space and the sense of the Earth as a planet had become familiar even before men first left footprints on the Moon.

The build-up to the IGY encouraged all the individual sciences involved to contemplate how they could achieve larger, more complex research endeavours. While there were enormous quantities of government money going into US science this was by no means evenly spread: Physics was rich on the Atomic Energy Commission; Biomedical Sciences getting fat on the National Institutes of Health, but for geologists there was no cornucopia of funds but instead the only Federal Agency, the US Geological Survey 'was small, conservative, and most damning of all, as far as academic science was concerned, did not give grants.'[8] At a meeting of the US National Science Foundation (NSF) in March 1957, eight scientists were gathered to review research proposals associated with the study of the Earth and at the end of two days and 60 proposals, one member of the Committee, Walter Munk, a renowned Californian geophysicist who had studied rotational and morphological properties of the whole-Earth, exclaimed (as quoted by Hess eight years later[9]) that 'none of these proposals was really fundamental to an understanding of the earth'. 'We should have projects in earth science – ', continued Munk '– geology, geophysics, geochemistry – which would arouse the imagination of the public, and which would attract more young men into our science.' Munk continued on this theme and then came up with just such a inspired proposal: 'to drill a hole through the crust of the earth'. Harry Hess who was also at this meeting endorsed this idea with enthusiasm. Thus was born Project Mohole.

The Mohorovičić discontinuity between crust and mantle as found from the evidence of the refraction of seismic waves, lies 25 km and deeper below the continents, but a mere 5 or 6 km below the ocean floor. While continental boreholes had already reached more than 6 km in depth, drilling below the oceans had barely pierced the sea bottom. The demand for such drill holes went right back to Charles Darwin's letter to Alexander Agassiz in search of some 'doubly-rich millionaire'. The wealth that Darwin needed to finance a drill hole through a Pacific coral island was possessed only by the military. In 1947 after the first test of the atomic bomb at Bikini Atoll a hole was abandoned at 779 m still within the coral. The three geologists involved tried again on Eniwetok Atoll and finally succeeded in proving Darwin correct, by finding volcanic basalt at 1287 m beneath the surface.

In the midst of the Second World War, T. A. Jaggar, president of the Hawaiian Academy of Sciences and founder of the Volcano Observatory on the rim of the great caldera on Kilauea, wrote a letter to Richard Field, proposing that 1000 holes be drilled to depths greater than 1000 ft (300 m), into the floors of the oceans of the world.[10] Jaggar and Field were both Harvard alumni, pupils of Agassiz's oceans and Pickering's volcanoes. Jaggar, a Harvard contemporary of the unique volcanic mobilist Daly, had even been an instructor to Field and a teaching contemporary at MIT. Following his compatriot William Lowthian Green, Jagger had become influenced by Kilauaea to contemplating a greater scope of investigations on a living planet. Viewed from Hawaii, Geology seemed a 'largely speculative' science based only on the continents. Jaggar estimated 20 million dollars would see the project started, that the geological community should team up with the oil industry to carry out the venture and that at the successful conclusion of the war, 'thousands' of ships and engineers would be idle, awaiting a new mission.

'Between the idea and the reality, falls the shadow'. Who should be put in charge of Mohole? Hess suggested the unofficial and irreverent American Miscellaneous Society (AMSOC) that had been founded in 1952 by two geophysicists at the Washington Office of the Naval Reserve (one of whom, Gordon G. Lill, had been involved in the atoll drilling programme) to provide a whimsical forum for ideas outside the scope of normal scientific discussion. A committee was set up to run the venture, among whom were Hess, Munk and Roger Revelle of Scripps. One week later Ewing was persuaded to join.

After an initial request for a $30000 feasibility study was turned down by the NSF, Hess set about making AMSOC a more reputable organisation, receiving unexpected assistance for the project from a Russian announcement that they already had the equipment for such an enterprise and were merely searching for a site. The original NSF proposal had suggested a single hole, like a single Moon shot. In articles written in *Science* and *Scientific American* in 1959 more detail was given, both of the anticipated cost of the venture ($5 million) and a wider vision of the project's scope: 'No one site or hole will satisfy the requirements'. Public interest had by now been successfully aroused. NSF awarded $80000 to Woods Hole and Lamont for drilling site surveys, and AMSOC came up with a new sum of $14 million for the whole venture. NSF promised $1250000 to allow the commencement of a four stage drilling programme. In 1961 permission was obtained to use a war-surplus Navy barge and at a location close to the island of Guadeloupe, off Mexico, in a depth of 3.5 km of water, a hole was bored 200 m into the ocean floor. At the bottom of the core was basalt. Hess was delighted. So was President Kennedy, declaring the mission 'a remarkable achievement and an historic landmark'. Thereafter the programme fell apart.

The project was put out to tender and given to a firm of Houston engineers based in the congressional district of Albert Thomas, who was in charge of the allocation of the NSF budget. Many of the project's earlier advocates had lost interest in what was seeming more and more to be an elaborate and wholly symbolic rape of the mantle, rather than a realistic scientific research project. The anticipated budget estimate had risen as high as $125 million dollars, when after a scandal about the engineers' campaign contributions to the Democrats and the death of Albert Thomas, the whole project was cancelled by a Congressional subcommittee in 1966.

'In the entire postwar partnership of science and government, no research enterprise . . . produced greater political embarrassment, anguish, or defeat for the scientific community.'[8] As the feasibility studies continued, it became apparent that the original idea to drill a hole into the mantle was founded on little more than an elaborate hoax; a project whose central purpose was to deflect some of the imaginative impact of space exploration to more down-to-earth goals. The concept of the mantle had been invented by seismologists; the old substratum by definition was always concealed. The possibility that mantle material might occasionally turn up at the surface had not even been suggested, although Hess had already guessed that the peridotites on which he had done much of his life's research were actually samples of the most abundant of all Earth rocks – those of the mantle. While such a project remains an important scientific goal, reaching for the drilling rods at the first hint of ignorance would be as if an astronomer lamented that his subject was impossible unless some samples of star plasma could be bottled and returned.

There were a number of important spin-offs from the Mohole Project. The success of the first deep-sea drill in 1961 indicated that a far more reasonable goal would be the collection of data from shallow boreholes drilled into the floors of the world's oceans. In 1965 a series of demonstration holes were drilled and in 1967, after the certification of the death of Mohole, NSF put up the first instalment of a generous supply of money ($68 million by 1975) to mount a deep-sea drilling programme. The company that gained the contract, Global Marine, through the success of the technology developed for this enterprise, came to the attention of the CIA and in 1970 a sister ship to the original deep-sea drilling project's *Glomar Challenger* was undergoing modifications and tests, under a contract arranged through Howard Hughes, to attempt to lift a Soviet submarine off the floor of the Pacific Ocean. After paying Hughes a total of $350 million dollars (five times the scientific research budget) on this unsuccessful lifting operation, the CIA abandoned the mission when the cover for the operation was blown in 1975.

The least tangible, but perhaps most significant product of the ill-conceived and ill-fated Mohole Project was that it fulfilled its ulterior intention: to capture headlines for the Earth Sciences. The universal appeal of its goal, a goal in the spirit of heroic Victorian engineering, could unite all those who were to become Earth scientists in the spirit of this adventure. More subtly, for the first time geologists were offered access to the ocean crust that had previously been the preserve of geophysicists. Thus the phantom Mohole helped create through its fictions a genuine substantial new science; for soon after the first announcements of plate tectonics, the phantom had dissolved to pass on Earth Sciences to be blessed by the new global tectonics.

From the publicity of Mohole and from Wilson's experience of the IGY, it can be found that the creation of the Earth Sciences preceded the discoveries of sea-floor spreading and plate tectonics. The desire for departments to change their names also preceded this break-through; the same desire as expressed by Munk in 1957, to make the subject appear more glamorous. In the competition for students it was important to distance this new exciting science of the Earth from

'Geology', a name that was laden with dusty, antediluvian associations. These moves to transform the appearance and shape of the science were taking place solely in North America.

As the Academy of Sciences had early on set up an Earth Sciences division, so the name-changing was a response to the possibilities of obtaining more research money, and more students. Unlike the old Geology, the new sciences were hungry for funds, to support costly equipment both in the laboratory and in the field, as well as travel expenses to finance the globe-trotting needed to collect samples for studies of rock magnetism or age-dating. A global perspective had been assisted by widespread commercial air-travel. In place of mapping one area in detail, samples could now be collected from rocks scattered across the islands and continents of the globe. In the 19th century such a global view required years of voyaging, as with Darwin in the *Beagle*. The same scope could now be gained in weeks.

Several entirely new sciences had grown up since the war: there were the isotope geochemists investigating age-dating, the geomagnetists, the geochemists performing rock analysis to understand the chemistry and chemical pathways of global systems. Roubault, in his introduction to 'Sciences de la Terre' in 1953 had written of the need to break free from the traditional disciplines and welcome the newborn sciences such as Geochemistry.[3] The exploration of the geology of the oceans was being undertaken by geophysicists, not geologists; and geophysicists were being encouraged by the oil companies to improve their equipment and techniques, so to be better able to map the hidden rock formations. For a time these new sciences could exist in a Department structured around classical Geology, or chemistry, or physics but it was inevitable that this structure of organisation was going to change. The new sciences had ever-increasing momentum; some of the research efforts had been direct spin-offs from the Manhattan project and in the immediate post-war period had attracted large numbers of high-flying chemists and physicists to work on problems in Geology. If there was one feature that united all these diverse new sciences it was, as stated by Los Angeles's pre-eminent isotope geochemist, that 'there are no old men practicing the art.'[11]

A cultural revolution

Hallam, in his book *A revolution in the Earth Sciences*, offers a final, short chapter 'Reflections on the revolution' that begins: 'The title of this book contains an ambiguity. Did the "revolution" in the Earth Sciences commence with the formulation of the hypothesis of continental drift early this century or did it take place when it became more appropriate to speak of plate tectonics'. Having provided the reader with only one serious option (the other would be as if the Russian Revolution was dated at the year of publication of *Das Kapital*), Hallam finds confirmation for the existence of a 'plate-tectonic revolution' in the astounding speed with which the transformation of ideas took place; as though this speed confirms that it was a truly revolutionary revolution.

Yet the 'revolution in the Earth Sciences' was neither totally within the Earth Sciences, because they could not be said fully to be constituted, nor was it in the *construction* of the Earth Sciences, because the reconstruction had already begun. It is often difficult to unravel the usage of the concept 'Earth Sciences' in the 1950s exactly because those who recount it now speak in retrospect from the 1960s, by which time the phrase has gained such currency that it is employed with unnoticed abandon on all those, at all historical periods, who with the facility of time-travel could be placed in a modern Earth Sciences Department. This desire to rewrite history is so strong that even rigorous historians of science, like Roy Porter, were carried away by the enthusiasm of their contemporaries, anachronously naming natural philosophers, as far back as the late 17th-century foundation of the Royal Society of London, 'Earth scientists'.

To consider anyone who has ever pondered on, or struck a rock in curiosity, an Earth scientist is to sustain the same deceit that exists within the term a 'revolution in the Earth Sciences'. It is necessary for all institutions to insist on their inevitability. It was expedient for the new Earth Sciences to claim everybody, geologists, geophysicists, palaeontologists *et al.*, as forming a common part of their historical lineage and validation. That perhaps it is possible to identify certain individuals who did, before the founding of Earth Sciences, share its specific intellectual methods and objectives, requires some historical analysis and definition of the competing sciences of the Earth.

Geology has always been the study of the local earth's surface, through observations made on the rocks, through the collection of samples and the construction of maps. Geology was concerned with unravelling the history of the earth's land surface, in particular through analogies with present day processes. The backbone of Geology was stratigraphy and fossil correlation. From the mid-19th century the most important problems in Geology concerned the origin of mountain ranges.

In contrast, Earth Science begins with the study of the Earth as a planet as revealed by Geochemistry and Geophysics, and then moves down in scale to the particular problem under investigation. The evidence from rocks that can be collected is given equal status with indirect geophysical and geochemical evidence of materials and processes that cannot be reached. It concerns itself equally with the rocks beneath the sea and those on land, and those known to exist buried at depth equally with those found at the surface. The tools of isotope Geochemistry are used to find history out of rocks and minerals that can extend back in time to before the formation of the solar system, and to this end it employs an absolute timescale based on radio-isotope dating. It is concerned with the Earth as process rather than rocks to be collected and classified.

Geophysics became consolidated in the late 19th century, as an application of physics to the natural world, a science as new and as dynamic as its parent German nation. In Germany, Geophysics was Earth Science: expansive, global, and like Geology, a culture, enriched with history, debate and enthusiasm. At the launch of his *Beiträge zur Geophysik* in 1887 Professor Dr Georg Gerland occupied fifty pages in locating geophysics, the product of a union between his own Geography and the new Astrophysics: a lineage distinct from Geology, traced back to Kant and Alexander von Humboldt, the global German geographers at the beginning of the 19th century. Meteorology, hydrology, physical oceanography and Earth physics all lay within the span of Geophysics. The problem outlined in the title of Wegener's 'Origin of the continents and oceans' reposed precisely within the global-geographical frame of German geophysics as described by Gerland. Nowhere else in the world was there a culture that could discuss such questions.

German geophysics survived the First World War, despite losing to France the territory (at Strasburg) containing both the University base of the primary journal *Beiträge zur Geophysik* (that took a decade to recover from the expulsion) and the headquarters of the International Seismological Association. Even in 1930 there were more geophysical institutes in Germany than in the rest of the world.

British geophysics was never a culture; the demarcation of science between Physics, Geology, Meteorology and Astronomy discouraged any re-allocation of the natural sciences. The government Meteorological Office initiated 'Geophysical Memoirs' in 1911, but the content remained largely atmospheric. In 1916 a British Association committee was convened to discuss the future of 'those sciences for which worldwide observations are important', reporting in 1919 that a Geophysical Institute was required. By 1920 after Cambridge University had failed to raise sufficient funds and as British seismology's rapid decline had removed the main inspiration for such a body, the proposal lapsed. The Society for Geological Physics formed in 1916 survived only three years before losing its autonomy to a committee dominated by the Royal Astronomical Society. The establishment had successfully blocked the formation of a new science.

In the USA, like Britain, geophysics was seen as a series of unrelated techniques. The most vigorous geophysical research began with global magnetic observations at the Department of Terrestrial Magnetism, founded from the Carnegie fortune, in Washington in 1904. Unlike Britain, however, the move to found a geophysical society succeeded with the creation of the American Geophysical Union in 1919. Over the next two decades the AGU thrived on the new commercial interest in geophysics within petroleum prospecting and thus never gained the Humboldtian *weltanschauung* of its German precursor. It was the revolution in the Earth Sciences that finally gave American Geophysics a culture.

At the time of the Revolution in the Earth Sciences, there was also a

Revolution in the Campus and a Revolution in the Classroom. At the time of the Revolution in the Earth Sciences students of history and social sciences were being made familiar with the writings of Thomas Kuhn. Kuhn had been a Physics graduate in the early 1950s when he had become self-conscious about the sudden shifts in scientific belief. This self-consciousness led him to investigate scientific history, in particular that golden age of the 17th century when the cast list of science was so slim, and so to write *The structure of scientific revolutions*. In this book Kuhn coined the loosely defined, but nonetheless powerful, concept of the paradigm: the matrix of scientific theories that are held to contain one science's understanding of the world under study and the nature of the problems to be solved.

Science, according to Kuhn, does not show continual evolution, but sudden bursts of transformation – the scientific revolution. At such times one paradigm becomes replaced by another. The innate conservatism of scientists ensures that those who supported the fossil or redundant paradigm will often fail to become converted, just as they resisted the transformation before the revolution. Earth scientists of the 1960s reading Kuhn found that he spoke to them, unlike other philosophers of science, about concepts that they saw had immediate relevance to their studies. Within the world-shattering changes of belief taking place inside the revolution, Kuhn's model could even possess more truth than that to be found in science.

As far back as Bragg in 1922 there were prophets proclaiming that continental drift would cause a transformation as radical as that of Copernicus. Yet how would the revolution begin? Wilson has claimed that he was inspired to deliver his 1963 incitement to revolution through reading Thomas Kuhn (published 1962). Beyond self-confirming prophecy he had followed a manual on how to prepare a scientific revolution, changing the name of his department to Earth Sciences as he plotted the overthrow of the old order. Having borrowed Kuhn's historical mirror to divert the scientific future, it is small wonder the revolution in the Earth Sciences is beset with internal reflections. Did the revolution fit the model or the model the revolution? Most of those who have written on the revolution have introduced Kuhn into their arguments, yet the simplicity and success of his model has camouflaged the complexity of the actual transformation. The urge to reject the way in which Geology comprehended the Earth had been articulated by the new breed of geophysicist in the 1930s. This spirit of rebellion was at first formless; there was no suggestion of a cogent alternative. In the late 1950s the concept of Earth Sciences was created but still lacked a model for the behaviour of the whole-Earth that would turn an ideal into a coherent structure. Only in 1963 did Wilson show how continental drift validated the Earth Sciences and therefore why the revolution was now inevitable. All these various stages were integral to the eventual transformation.

As long as the revolution was claimed to be plate tectonics there remained a mystery: why hadn't it happened a generation or two before; why had Wegener not gained articulate converts who could have overcome Geology's prejudice? Hess's seafloor spreading was no more than his mentor Vening Meinesz's convection or Holmes' widening oceans, proposed thirty years earlier. Yet the core of this revolution was neither moving continents, nor a convecting Mantle but the Idea of the Earth. In Kuhnian terms there had been a paradigm shift from Geology's little earth to the whole Earth of the Earth Sciences. After 1970 dissertations, articles and books did not have to mention plate tectonics to be revolutionary; they just had to contemplate the hidden and visible Earth at a whole new scientific scale. No longer would it be possible for errant apocalyptic cosmologies, such as those of Horbiger or Velikovsky, to rampage across the deserted territory of whole-Earth theories because finally that vast tract had been colonised by Science. A simple and unexpected 'revolution' as described by Hallam would leave a structural vacuum; Earth Sciences had been existing and growing in parallel with Geology for almost a decade. That this parallel structure existed was more important to the success of a revolution than the eventual trigger: the announcement of plate tectonics.

An alternative history of plate tectonics

And where in the flow of ideas did plate tectonics come from; was it the child, or perhaps the grandchild, of continental drift? It is a common and mischievous deceit on the part of detective story writers to follow the actions of certain characters and thereby simply by weight of detail to cast them as chief suspect. Let us gather the various interested parties into the drawing-room and assess the evidence.

Continental drift was a wild, fantastic idea; 'the striking similarity of coast lines between Africa and Brazil . . . must have been "made by Satan"'.[12] It had its origins in the desire in the 19th century to construct new myths, half in and half outside science; myths that would explain the paradox of the Americas; that would place the birth of the Moon in the context of the figure of the Earth. Even for Wegener the distorting influence of Greenland was mythic in form. When continental drift gained for its brief period of ascendancy a popular reception, it was upon an emotional appeal; it could never hope to convince those who sought to test it against scientific criteria, because like all whole-Earth theories of that period, there were no measurements to be made; just analogies, impressions and argument. It was a pseudo-scientific hypothesis; kept afloat only by the hot air of discussion.

When, in 1944, Bailey Willis wrote contemptuously 'Continental drift: ein Märchen', he was not to know that this story was going to have a real-life happy ending. That continental drift, after sinking into obscurity, should suddenly in the 1960s become acknowledged as being a correct explanation for the Earth's behaviour, was a fairy-tale, wandered far from its natural habitat into the world of science. This fairy-tale is so captivating, so much part of every Western child's understanding of the fictional structure, that it has been repeated by all the plate tectonic story-tellers in much the same way. Continental drift appeared, was dismissed, required more verification, and then reappeared. Despite the fact that she has no coach, gown or slippers, Cinderella does go to the ball. For a brief spell she has the attention of all the guests, for she is radiant. She gets to dance with the Prince of Science. But when at midnight her costume is revealed to be insubstantial; no more than rags, she returns to hearthside obscurity, to be mocked and ignored by her ugly elder sisters. Yet at last the glass slipper that she left behind, the little fragment of magical truth, is found to fit her alone; her sisters denying or renouncing their former hostility, she returns to marry the Prince and live in splendour.

Little Cinderella continental drift is very pretty and the story of her social climbing very insistent. But is it correct? Could we, for example, write a history of plate tectonics that entirely ignored Alfred Wegener? For the journey from continental drift to plate tectonics is very much the view from Geology. As long as the continental drift debate was fought along geological lines, arguing for analogies between mountain ranges, fossils and climates to either side of an ocean it was always to be inconclusive. There are two histories that led to plate tectonics: that of continental drift was the great inspirational Myth of the Earth.

However, it is a second invisible history that is arguably the more important. It is a mixture of globalism and the collection of scientific facts of a form beyond the descriptive. For plate tectonics is a 'hard' scientific theory, the antithesis of the conceptually 'soft' continental drift. In searching for the origin of continental drift in the 19th century, it proved possible to list a series of clues that were awaiting assembly. The same is possible for plate tectonics, only the two lists do not overlap. This second history provides a separate ancestry to the Earth Sciences, born out of a scientific approach to the Earth. The two figures who stand out as being true pre-emptors of Earth Science in the 19th century, Dutton and Fisher, were both trained outside Geology and both sought neither to subjugate the Earth to the requirements of physical models, as did the Cambridge mathematicians, nor to exclude Physics from the behaviour of the Earth, as did the geologists. Arthur Holmes in the 20th century was originally a student of Physics and had the scope of an Earth scientist but failed to perceive the fundamental subjectivity of Geology; for having provided a brilliant analysis of the importance of convection to continental drift he allowed the theory to languish in obscurity and subsequently abandoned this line of research, claiming when he was over 60 in 1953 that 'despite appearances to the contrary, I have never succeeded in freeing myself from a nagging prejudice against continental drift; in my geological bones, so to speak, I feel the hypothesis to be a fantastic one.'[13] Samuel Warren Carey had also been a student of Physics before becoming a professional geologist but he too became diverted by the 'geologists duty' in refusing to contemplate the physical significance of his Expanding Earth.

Some of the most important 19th-century statements that suggest a premature conception of an Earth Science exist only as shadows. The journey from Elie de Beaumont's Earth to the new geometry of plate tectonics is the history of the struggle to escape from the imprisonment of Platonic and Euclidean geometry. The geometry of lines and planes that had its rigid Classical framework imposed on the imagination of schoolchildren, was the same geometry that Kant believed was implicit in all men. It took a great geometer, like Euler or Gauss, to conceive of the geometry on the surface of a sphere. This is a true revolution in conceptual thought. Why else in 1965, as he proposed both transform faults and plates, did Wilson not see that movement on a sphere imposed specific and easily demonstrable geometric constraints? Flat maps, Mercator's projection; all these restrained the truth – that we live on the skin of a globe and that the geometry of any horizontal movement of the rigid material of the outer shell of the Earth must follow small circles around a Pole as implicitly as a train moves on its tracks. Where the rigid ocean crust moves down into the mantle at a subduction zone, the curve of the ocean trench and the neighbouring arc of island volcanoes is a simple geometrical constraint of the intersection of two pieces of spherical shell, like a dimple placed in a ping-pong ball. The ocean lithosphere is denied the possibility geometrically of assuming any other shape. All this geometry was there awaiting discovery. While the island and mountain arcs were noticed by Suess it was Sollas in 1903[14] who perceived (without explanation) how exactly some of the circles are drawn.

Yet the Earth only exhibits geometry when viewed as process. William Lowthian Green remarked that 'Volcanoes . . . cannot be regarded as local and accidental features, but must be contemplated together as a grand cosmical phenomenon' (see page 22, Chapter 2). He also observed presciently that 'movement of the earth's crust, is in the broad sense, volcanic action, and it is an action which has left its own record.' The 'record' was the moving ocean crust. Almost 100 years before Wilson had understood the profound significance of the 'grand cosmical phenomenon' of the Hawaiian islands, Green had arrived at almost the same conclusion, seeing the regularly spaced islands increasing in age and sinking beneath the ocean as they passed to the north-west: 'the Maui group of islands appears to represent the Island of Hawaii partially submerged.'

While the relation between volcanoes and plates is not always simple, the plates were defined on one set of observations alone: the global pattern of seismicity. The science that contributed directly to plate tectonics, without passing through any mediation of Geology or continental drift was perhaps the most neglected of all the Earth Sciences, the study of the distribution of the world's earthquakes.

After Lyell, geologists shunned the study of earthquakes as if they offered nothing but a promise of disruption and disarray to the new

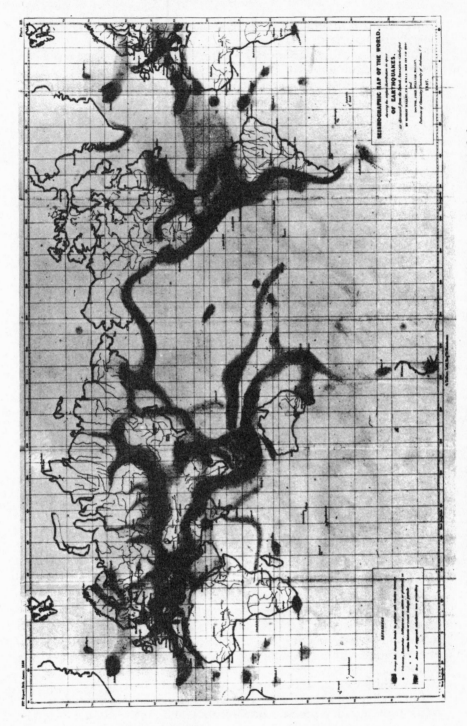

Mallet's global earthquake map of 1857 – all but the mid-oceanic plate boundaries are revealed

order of Geology. For an earthquake, unlike its cousin the volcano, provides little that can be collected and classified. Even after the awakening of the science of Seismology in the 1880s earthquakes were considered to be a branch of Physics or Geography, not Geology. The restriction of earthquakes to particular zones of seismicity and the global pattern of these zones was first revealed in all its details on a map prepared in 1857 by an Irish civil engineer turned seismologist: Robert Mallet. His map was not to be improved upon until after the Second World War.

Earthquakes only began to interest Mallet after he was 35 years old. From 1850 he compiled a catalogue of all the earthquakes that had ever been recorded; a list itself based on many earlier catalogues from which in 1858, he produced a giant world map, measuring 75 inches by 48 (190 × 121 cm), of earthquakes coloured in three tints according to their severity.[15] His ideas as to the origin of earthquakes were progressive: 'by the sudden flexure and constraint of the elastic materials forming a portion of the earth's crust'. However, he came to see earthquakes as being typical of the Earth's behaviour and volcanoes as aberrations. In a paper read before the Royal Society of London in 1872, he showed that as earthquakes were mechanical phenomena, so too were volcanoes produced locally by transformations of the mechanical work of contraction through compression or crushing.[16] Like Elie de Beaumont, he was in the parlance of that age, a great telluric engineer. 'It is given to no man', wrote Mallet, with mitigating modesty, 'so to interpret nature that his enunciation of her secrets shall remain for ever unmodified by the labours of his successors.'[17]

Mallet's Earth science was continued by his near-contemporary at Trinity College, Dublin, Thomas Oldham, who after studying engineering and geology at Edinburgh, returned to Dublin becoming professor of Engineering and director of the Irish division of the UK Geological Survey. In 1850 Oldham travelled to India to found the Indian Geological Survey, investigating the Cachar earthquake of 1869 according to Mallet's methods, as well as documenting the history of Indian earthquakes. Thomas Oldham's son Richard (born in 1858) completed his father's earthquake work, and himself became superintendent of the Indian Geological Survey. His detailed study of the extraordinary damage and ground deformation associated with the Great Assam earthquake of 1897, led him to utilise distant seismic records collected by Milne. He identified the three types of waves present in these records and thus, after his return to England in 1903, wrote of the existence of the Earth's core (see p. 64).

Richard Dixon Oldham was the last British earthquake geologist, heir to Mallet's vision. For the years 1920–2 he was President of the Geological Society of London, where he attempted to infiltrate some physics, inviting lectures from the astronomer Arthur Eddington (see p. 74) and from a physicist, C. G. Knott, talking on seismological investigations of the Earth's interior. Oldham introduced Knott almost apologetically, reminding the audience of the wartime need to fill the supply of papers with topics outside Geology. Oldham had retired before the same argument could be used to justify a debate on Wegener's drift.

The scale of continental drift and plate tectonics prevented either phenomenon from being visible, yet, although these theories could not be seen, they could be evidenced in two antipodal natural media: ice and molten magma. Those such as Dutton, Powell and Gilbert who believed in a plastic substratum always employed the analogy with ice. Anyone who has crawled inside a crevasse, marvelled at the seracs in an ice-fall, seen how ice is brittle but also how it can flow, how a glacier is no more than a great tongue of liquid hanging off the mountainside; anyone who has experienced ice over a period of time is left with a deep-seated impression of solidity and flux bound up together. That impression is more powerful than any physical description of the properties of the mantle. Wegener had spent 18 months living on the edge of the Greenland ice-cap before returning to Germany to experience satori. Wegener's continents were ice-floes but his substratum had the properties of the ice-cap.

And yet there is one problem with seeing the behaviour of the outer shell of the Earth within the world of ice, and that is that ice always floats on the water and thus nothing in the visible world of the Arctic can give evidence of overturning and convection. One of Wegener's companions on his first Greenland expedition had claimed that continental drift had arrived through observations of ice-floes. This certainly explains the limitations of Wegener's perception of drift.

These new images obtained from exploration in the polar regions came to overlay an earlier understanding of the behaviour of the Earth to be found in the writings of the 19th-century fluidists. Continental drift and the solid Earth of the seismologists effectively ended this tradition, despite the attempt by Holmes to integrate the older fluidism in with moving continents. Through their concern with motion in the substratum, the fluidists were often closer to the conception of the modern Earth than many drifters who treated the underlying substance of the Earth with a geologist's disdain. The fluidists, Dutton, Fisher and Green, all gave great prominence in their Earth models to observations made at a single location: the magma lakes high on the flanks of Kilauea on Hawaii.

While Green was the watchman for these volcanoes, they also provided a suitable destination for many influential 19th-century scientists. The American geologist Dana went to inspect Kilauea and Mauna Loa, with the US Exploring Expedition. Dutton spent much time watching the formation and disruption of a solid lava crust on top of pools of incandescent magma, but even Dutton's descriptive powers were surpassed by those of another distinguished American traveller, who came in 1866 to admire these curiosities. Mark Twain, like Green, had arrived from California after dropping in on the Gold Rush.[18] He climbed up the flanks of Kilauea, then in eruption, and was spellbound by the size of its crater:

Here was a yawning pit upon whose floor the armies of Russia could camp, and have room to spare . . . over a mile square of it was ringed and streaked and striped with a thousand branching streams of liquid and gorgeously brilliant fire! Occasionally the molten lava flowing under the superincumbent crust broke through – split a dazzling streak, from five hundred to a thousand feet long, like a sudden flash of lightning, and then acre after acre of the cold lava parted into fragments, turned up edgewise like cakes of ice when a great river breaks up, plunged downwards, and were swallowed in the crimson cauldron.

Mark Twain had just witnessed mid-ocean spreading ridges and sea-floor subduction.

These observations became quoted by Fisher in his *Physics of the Earth's crust*, along with Dutton's own descriptions of a more alarming form of catastrophist tectonics:

It is an exhilarating sight to stand at night upon the brink, and watch the boiling, surging, and swirling, of six acres of melted lava. At brief intervals the surface darkens over by the formation of a black solid crust, with streaks of fire around the edges. Suddenly a network of cracks shoots through the entire crust, and the fragments turn down edgewise and sink; leaving the pool one glowing expanse of exactly the appearance of so much melted cast iron'.[19]

Fisher provides these descriptions in order to justify his belief in a layer of molten rock between the crust and the solid inner Earth. If the lava of the crust was never lighter than the magma then why should it ever float?

Fisher even went one stage further, taking this behaviour of the Kilauea lava lakes to model whole oceans. The upwelling substratum rises along the mid-ocean plateau: 'It is on these plateaux that the volcanic islands of midocean are based'. At the edge of the oceans, like the diving lava-floes of Kilauea, Fisher claimed 'most of the earth-quakes which disturb the much shaken islands of Japan originate beneath the sea on that side; which shows that the sub-oceanic crust in that region is in a very unstable condition, as it would be if it were thus sinking.' Although Fisher's oceans opened with the loss of the Moon and his convection did not involve the outer crust here we have the complete motor of plate tectonics without which continental drift was always so impotent.

At the revolution in Earth Sciences the Kilauea lava lakes, so familiar to Dana, Dutton, Green and Fisher, had to be discovered afresh. In the winter of 1970–1 Wendell A. Duffield of the Hawaiian Volcanological Observatory photographed and filmed the motions of the lava-floes within a small crater on the flanks of Kilauea. Such behaviour had been continuing through the century since Twain's visit, but only at the arrival of plate tectonics did it have sufficient meaning to be studied. Duffield observed 'a naturally occurring miniature version' of sea-floor spreading. The incandescent magma rose along lines 1 m wide and cooled into a black basalt, which some distance from the spreading ridge would plunge back down into the magma pond. 'The principal plate-bounding structures – rises, sinks and transform faults – were all clearly displayed on the laval column in an impressive, ever-changing mosaic.' The film and photographs were sent around the world so that plate tectonics could for the first time be visible.[20] It came as a total surprise that a phenomenon so radical and new could be happening in such a normal fashion, unnoticed.

The 19th-century fluidists' greatest legacy was to have directed their attention to the heat pouring out through the ground. Fisher begins his book *Physics of the Earth's crust*: 'There is no fact more firmly established in terrestrial physics than that the temperature of the rocks of the earth's surface increases with increasing depth.' Joly, Daly, Holmes, Bullard, all obtained their picture of the Earth at depth primarily from heat. The 19th-century mathematical physicists could not countenance a convecting interior because it would make their simple calculations of the age and behaviour of the Earth impossible. In 1929 Jeffreys chose not to grapple with Holmes's new advocacy of convection, and in the 3rd and 4th editions of *The Earth* in 1952 and 1959 prefers to believe that the material of the Earth is still too strong. The widespread gravity anomalies, interpreted in the earlier editions as indicating strength at depths in the Earth, were becoming explained by others as density differences in the convecting system. Of all these matters, Jeffreys remarks only that 'the suggestion of systematic convection currents complicates the question'. Fixism, as Argand recognised, implies a rigidity of mind. That there still remain problems with the detailed calculation of the physics of the convecting system is no longer considered a reason for demanding that the Earth behaves according to a simpler textbook.

The leaders of the revolution

The history of the rise of sea-floor spreading and plate tectonics initially escapes simple comprehension because of the sheer number of characters involved. Yet the number of combatants was, like some opera, a product of the story's own medium. While Galileo could observe all the planets and moons of the solar system with his newly acquired telescope, so the investigation of the whole-Earth that eluded man's perception required the collective vision of many individuals. The telescope of the oceans from dome to lens was in effect a whole navy, for oceanography had had a central role in strategic operations since Galileo's time.[21] The telescope of seismicity was a network of 120 WWSSN stations. The theory of plate tectonics was largely paid for by the US Defense Department.

Yet who, and many a bar-room gathering of geologists and geo-physicists has seen fit to discuss the topic, would, even should, win a Nobel Prize? If theoretical astronomers can award one another this curious product of an earlier Big Bang, then why not those who mapped the theoretical behaviour of a globe of matter, midway in size between the atoms and the stars?

First Wilson, the marine geophysicist who never went to sea. While Field had trained Bullard, Hess and Ewing to go out into the oceans and collect the data, he failed to so educate Wilson, but instead inadvertently moulded him in his own image; as a carnivore for the data of others. Was Wilson therefore 'the Newton of drift'? If Wilson had worked with his former researchers, Scheidegger and Morley, then plate tectonics would have been a purely Canadian discovery. There was no one man who saw it all the way through; was the journey too arduous and strange? Together Wilson and Vine were the leaders in the revolution; but if there are prizes to be awarded then they should at least include a large number of certificates of merit.

This choice is fortunate because the awards might otherwise be posthumous. Of those who had been involved in the great debates of the 1920s only Sir Harold Jeffreys lives on. Holmes, who had taken no part in the reawakening of drift, died in 1965. Vening-Meinesz who had spoken at the Royal Society Conference in 1963 and who therefore did bridge the two great phases of debate, died in 1966. Harry Hess, the lifelong chain-smoker, died of a heart attack in 1969 while chairing a committee of the Space Science Board of the National Academy of Sciences. In 1972 after the student riots at Columbia, the university found itself in financial straits, and Ewing believed that the new President William McGill desired not only a greater control over the activities at Lamont but also a part of Lamont's Doherty estate. The dispute flared, and Ewing, a year from retirement as Director, resigned with only a month's notice, returning to his home state of Texas, and taking with him a number of his closest colleagues. He died of a cerebral haemorrhage on 4 May 1974. Within months a book[22] had appeared, first serialised in the *New Yorker*, which was to celebrate Ewing not only as the rugged all-American founder of the greatest oceanographic research institute, but also, in the manner of a hagiography, to bind him to the history of plate tectonics in such a manner that a casual reader would be unlikely to believe anything other than that Maurice Ewing and Lamont alone had been the driving force behind the battle to create a new mobilist global Geology.

Bruce Heezen, the dynamic and effusive cartographer and cartoonist of the ocean floor, abandoned his expanding earth in 1966 and became an enthusiastic sea-floor spreader. About the same time he had a dispute with Ewing that cost him his ready access to shipboard time and departmental space. He died 'in action,' of a heart attack, while aboard a diving submarine in June 1977.

Sir Edward Bullard was the most generous and creative of Field's students. While Ewing was driven by a hunger to collect more and more marine data, and Wilson to assess the whole pattern of information from across the globe, Bullard was the greater polymath, advancing both the methods and the equipment, viewing whole-Earth theories with an open mind until he could see that the argument had tilted in favour of mobilism. Bullard was also an administrator and a teacher. His success does not just include his own output but that of the small department that he led in Cambridge, and its students, that included Vine, Matthews and McKenzie. Through his custom of inviting colleagues from all over the world to spend their leave at Madingley Rise, he precipitated the formative collaboration of Hess, Wilson and Vine.

Bullard's most quoted quotation is 'All scientists have two homes; their own and the United States'.[23] His love for America had begun with his first visit to see Field. In the aftermath of the Second World War he saw that the funds flooding into science in America made that country the true patron of discovery. He had held a visiting professorship at Scripps since 1966 and it was there that he died after a long battle with cancer on 3 April 1980. He admired above all other scientists, Edmund Halley; like Halley, Bullard was an adventurer, undertaking fieldwork around the world, combining both the skills of a craftsman with the imagination of an artist. At Scripps the two sides of Bullard are remembered: the frivolous Bullard, the backside of Bullard, illustrated in a wall-mounted photograph showing him seated on a boulder, naked apart from a sombrero; and Bullard the scientist, in an obituary written by his friend and co-researcher Roger Revelle, emphasising 'Depth of Intelligence', 'Nobility of Character':[23] these were the affirmations given to celebrate this great geophysicist.

9

The eclipse of Geology

The reaction of the geologists

In the drift debates of the 1920s it was asserted that 'the surest test of drift's validity lies in the domain of geology'.[1] By 1964 it was admitted that 'purely geological evidence cannot prove or disprove drift'.[2] In reconsidering the earlier fate of Wegener's theory it became common-place to state that it was Jeffreys who killed drift; but his opinions had only provided confirmation for a deeper distrust that had prevented all geologists in America and all but a few in Britain even considering the theory worthy of discussion. The resistance of Geology to alien whole-Earth theories was primal, for the ultimate success of a theory based on geophysical information would involve the loss of Geology's traditional authority. Du Toit had recognised how complete would be the restructuring of the science if such whole-Earth theories gained ascendancy. Rollin Chamberlin had perceptively claimed that if Wegener's theories were correct, 70 years of investigation would be as nothing. The scale had shifted from the mountain to the planet; from an unrelated set of historical reconstructions based on rocks to the unification of global tectonics.

After plate tectonics one reaction of the geological establishment has therefore been to deny that there has been any revolution. Most significant of such revisions was a paper entitled 'A defense of an "Old Global Tectonics"' written by four geologists from the American South in 1974.[3] Their mission is well stated: 'It is our premise that this appellation (new global tectonics) is an unfortunate one in that it misleads laymen into assuming that the startling advances in our understanding of the earth are truly "new" or revolutionary'. They proceed to find precedent in the same figures discussed in this book. From this priority they wish to prove that nothing significant has changed: 'Our basic reason for writing this paper is simply to document briefly the extent to which our geologist predecessors had anticipated major aspects of plate tectonics.'

The view from Earth Sciences has shown that much of the

geologists' large-scale researches was misguided. The pursuit of the elusive geosyncline was now over; no longer the central mystery of the science, the term 'geosyncline' was itself discontinued. Thus while Geology as the study of the surface rocks will continue, it is no longer the whole Empire but now just one part of a much greater whole. The revolution involved a shift in the balance of power from Geology to Geophysics. That the death of the old Empire of Geology has not been announced, that there has been no obituary, is because of the conservative belief that Geology can accommodate itself to the new discoveries without need to alter its intellectual structure. One measure of this comes from observing the changing perception of Wegener among geologists and traditional Earth physicists.

In the early 1960s Jeffreys former student – the mathematical seismologist K. E. Bullen, in writing up Wegener for the *Dictionary of scientific biography*, and thereby giving Wegener his scientific location, chooses to condemn the man as a monomaniac, if necessary to mythologise him. Wegener is seen to have spent his life pursuing the single obsessive aim of proving continental drift. Moon craters, Greenland exploration, Meteorology, Climatology, all these vanish in the face of the moving continents: 'His search for geodetic support was one of the main motives for his third and fourth Greenland expeditions' claimed Bullen, 'In his endeavours to test his continental drift theory through geodetic measurements he designed an efficient balloon theodolite for tracking balloons sent up from ships to great heights'. Both statements are without foundation, the second is most extraordinary. How could Bullen possibly have believed that balloons floating in the upper atmosphere would reveal continental drift; unless he needed to presume that all Wegener's lunatic activities were centred around furthering the heresy? (In 1933 Baker had elaborated this myth into martyrdom: 'Professor Wegener most lamentably became a sacrifice to the theory of continental drift for he lost his life in Greenland seeking evidence of a continued movement of the continents'.[4]) Bullen continues: 'The enthusiasms of a considerable number of earth scientists lead them to assert, sometimes with a religious fervour, that continental drift is now established.' To introduce aspersions to 'religious fervour' into a scientific debate had connotations going back to Galileo.

In compiling the *Source book of Geology 1900–1950*, the editor K. F. Mather, writing in 1966, was happy to give space to Bucher and Chamberlin but excluded all mention of Wegener or drift; an editorial decision that only five years later he was to class as 'a very serious oversight indeed'.[5] This was, of course, no oversight, for Wegener was no geologist; yet for Geology to survive he must be reclaimed as such: Mather has been caught on the rebound, writing in 1971 that 'Wegener had one of the most fertile geological imaginations of this century'! Thus the need to locate Wegener as outside Geology becomes altered to the need to relocate Geology so as to include Wegener.

The most articulate opposition to plate tectonics in America was continued by Arthur A. Meyerhoff and his father Howard A. Meyerhoff of the American Association of Petroleum Geologists, who sustained, through the 1970s, a regular series of contributions listing all those geological observations contrary to the new theory.[6] (The father was publications manager of the American Association of Petroleum Geology and an emeritus professor of Geology.) A large part of their attack was dedicated to condemning Hess and Dietz for their irreverence in having ignored the work of earlier scientists. It is some sign of the collapse of Geology that there has been no wider opposition; for the Meyerhoff's attempt nothing more radical than to offer the model of the Earth that has survived at the centre of Geology unchanged from the 19th century. Thus their oceans are as old as the Earth, and the continents maintain their permanent positions. Their desire is, above all, to preserve the geological status quo. Thus much of their attack has the form of mitigation, of elaborate alternative explanations for magnetic anomalies, and a denial of all matching between the continents. Their most cogent arguments concern the rare discovery of older rocks in dredge hauls from the Atlantic and Indian Oceans that have been 'explained away' by the sea-floor spreaders as stones carried by drifting ice. To a geologist a rock will always be far more important than a magnetic anomaly.

Despite the protestations of the Meyerhoffs, even the average American geologist had become convinced of the plates; in a random sample survey carried out in 1978, 87% of them considered that the theory was 'reasonably well established'.[7] The existence of a Philosophy of Geology, the existence of an Earth Science, might be just as irrelevant to a prospector as the theory of evolution is to a fisherman. However, practical Geology of the 1950s and 1960s as carried out for engineering, mining and hydrocarbon exploration employed information from indirect geophysical techniques on equal terms with rock samples, and therefore has proved a great stimulus (even the economic motivation) for a transformation to Earth Sciences.

The main resistance to the belief that the old empire of Geology has passed away often comes from those geologists who were trained before Earth Sciences and who have sustained the old specialisations in universities and colleges. The revolution in the Earth Sciences only took place as a revolution amongst a group of young researchers in that least history-conscious of all societies: America. It was a genuine revolution because mobilist theories had previously been so effectively outlawed. These researchers switched allegiances, while in England those geologists who already had some sympathy for continental drift simply became confirmed in their beliefs.

The geologists of the old school of a fixed Earth cannot love plate tectonics, boosting young geophysicists with little knowledge of fieldwork to take command over older geologists. As late as 14 March 1967 Wilson received a letter from the director of a certain 'distinguished geological survey' claiming 'The opinion of the National Committee is that the subject of continental drift, attractive and stimulating as it is, is not of priority interest to geologists in (our survey)'.[8] John McPhee, writing in the *New Yorker* in 1982,[9] spends time in the company of a field-geologist who voices her intense disbelief in the plates. Other geologists are less eager to publicise their suspicions in the knowledge that there is a new orthodoxy. As television programme makers have learnt to their puzzlement, there is nothing in the whole of local geological investigation that provides a demonstration of plate tectonics. Plate tectonics cannot be seen through geological eyes.

In America the national Geological Society gradually became a forum for plate tectonics, yet it was the American Geophysical Union that naturally came to dominate the Earth Sciences. In Europe, although plate tectonics was generally, and for mainland Europe rather gradually, accepted, there was no immediate organisational restructuring of the science. The European influence even tended to dilute the American enthusiasms. Thus when Le Pichon had returned to join his colleagues at France's Brittany Oceanographic Centre at Brest, he published with them in 1972 a book *Plate tectonics* that was less than enthusiastic about attempts to explain mountain systems (the province of Geology) by the plate tectonic explanations of those such as Dewey. The Open University Department of Earth Sciences was an entirely new creation and therefore could be initiated with the new structure. Yet only 60 km away at Cambridge University, there existed three entirely separate departments which, after a decade of missed opportunities, were only forced by bureaucratic manoeuvring to fuse into Earth Sciences at the end of the 1970s.

In England and in Europe the revolution was personal; while in America it could hope to be cultural. In England, unlike America, there is still no central forum for the new geophysically based Earth Sciences, that first surfaced and continue to be most discussed at the Royal Society of London.

The indifference of institutional Geology can be measured by the neglect offered to the 'revolution' by the Geological Society of London. During the 1920s Presidents of this Society would allow the occasion of their annual address to consider greater issues of Geology and topics of current importance including drift. From 1964 to 1970, during the years of date tectonics, the presidents discussed: 'Lower Palaeozoic plankton', 'The problems and contribution of absolute chronology in Pleistocene stratigraphy', 'Practical geology and the natural environment of man: parts one and two' and 'The classification of Avonian limestones' with no mention of any of the new global tectonics.

At the crucial period there was no symposium held specifically to discuss plate tectonics. It was as if the inhabitants of some ancient building scrupulously ignored the erection of a glittering skyscraper on a neighbouring lot. The only occasions when geologists were privileged to glimpse the reconstruction taking place were when Tuzo Wilson delivered a short talk in 1965 without any discussion[10], and at the Annual General Meeting of April 1967 when Kingsley Dunham, Arthur Holmes's most illustrious protégé, presented Bullard with the Wollaston Medal. In his introduction Dunham apologised that 'for long it has seemed to me a matter of regret that, in the earlier years of this century, the incursions of physicists into earth science received a less than enthusiastic welcome from this Society'. Bullard generously thought that such sentiments might reflect 'irrational guilt'. He opens his reply with 'Sir, you and I are lucky men: in our lifetime we have seen our subject transformed from a backwater to a band-wagon.[11] How many of those in the audience knew to what he was referring?

The Geological Society of London had become a hostage to fortune, in a way that was not inevitable in the original constitution of Geology. For when in the early 19th century Geology was both popular and successful it was also a broad church – a science that could accommodate dissent, a science that could have grown into the Earth Sciences. For the first of the Earth physicists, William Hopkins, who established a much-echoed precedent in his complaint 'that he could not get geologists to understand mathematics nor mathematicians to take an interest in his geology'[12] proved such a challenge to the Geological Society that from 1851–3 he was made its President. Geologists had recognised how mathematics offered new insights into the Earth. Yet within a few years this greater Geology had been divided. A former secretary of the Geological Society, one Charles Robert Darwin, published a book that inexorably linked Geology with atheism and sent mathematicians scurrying to the support of the angels. The battle with the physicists over the age of the Earth, left the geologists, like any group that has suffered under an oppressor, imprinted with distrust for mathematics. The rise of the new Geophysics after the First World War could have been accommodated in Geology, just as Hopkins had been assimilated with a presidency. Instead the geophysicists were forced to work with the astronomers: from 1922 the Royal Astronomical Society initiated a geophysical section. From the day the geophysicists were made unwelcome, Geology inevitably drifted away from the centre-stage of Earth investigation. The rest is history. After 1970 the Society developed many sub-groups of specialists, the bold authority of Geology having become splintered and diffuse. After an episode of introspection in 1971, the society compiled a new motto, *'Quicquid subterra est'* which was set on an emblem dated 1807 and mounted on the cover of the year's supply of Proceedings as if the buried Earth had been at the heart of the Geological Society's enquiries since its foundation. At the same time, the parliamentary debating chamber furniture of the meeting room was torn out to make way for a lecture theatre. The old adversarial Geology in which the two sides of an argument (catastrophist/uniformitarian, fluidist/solidist, mobilist/fixist etc.) had no more objective merit than that to be found in a political debate, was to be replaced by the 'scientific' geography of speaker and audience. There have been occasional subsequent attempts to relive the days of the Geological Raj.

The Geological Society's neglect of plate tectonics during the 1960s finds its counter-point in the welcome offered to a rival whole-Earth theory at the end of the 1970s. In 1979 a meeting on 'The expanding Earth' was held at the Geological Society and publicised with press releases as if some important breakthrough was to be announced. Yet the meeting, launched by Carey, proved to be a close reproduction of the expanding earth debates of the 1950s. As in the steady-state creation theories of that period, the mass of the Earth was increasing but only under high pressure – hence there was little evidence for increase at the surface. A question was asked from the floor: could not an experiment be set up – perhaps two large masses suspended on a balance, one under high-pressure, and so detect slight changes in mass through time? The chairman declared that the discussion was getting too far outside Geology.[13] The meeting was adjourned.

The expanding Earth theory is not however just a historical curiosity – a halfway theory that the more precocious geophysicists, such as Bruce Heezen who believed in Carey's model for almost a decade, or J. T. Wilson who in 1960 briefly flirted with expansion, used as a stepping stone in their conversion to the plates. Its survival provides illumination on the cultural and philosophical methods of Geology, as being, as the chairman had identified, distinct from those of the physical sciences.

An important argument against the expanding Earth, an argument that could equally be employed against the contracting Earth, was stated unequivocally by Clarence Dutton[15] but comes from a theorem first propounded by Karl Friedrich Gauss concerning the property of surface curvature. Dutton provided a powerful analogy: those favouring a contracting Earth always had the model of a collapsing arch in their minds when they should have had a collapsing dome. Without considerable demolition, domes are hard to shrink, or to expand. Suppose the interior of the Earth expands, as Carey suggests, by 20% at an ever-increasing rate. The rigid continental crust that once covered the whole planet is arranged according to the original curvature. Although on a two-dimensional map it may seem possible to inflate the Earth and pull the continents apart, a simple test to lay large sections of orange peel on the skin of a football will demonstrate the difficulty. If the interior of the Earth had expanded then it would require that the continents had been torn apart down to a small scale to disrupt their original curvature to a point where it no longer mattered – the equivalent of cutting the orange peel into small pieces. This Gaussian curvature is well known to all mathematicians and many Earth scientists, but geologists who champion the expanding Earth have refused to reason with it because they believe, along with their 19th-century forebears, that Geology should never be constrained by mathematicians. In this instance this means that Geology can hope to transcend the very properties of space.

In parallel with the 19th-century contraction theory the expanding Earth is structurally compatible with the old empire of Geology. Not only does it claim that the Earth's interior is a mystery that can never be resolved because it can never be seen, but it makes traditional 'geological' demands on the behaviour of matter in this interior. Having won the battle with the physicists over Earth-heat, the moral of the victory is continually restated, as it was at this expanding Earth meeting: Geology proposes, Physics predisposes. It was fundamental to the self-opinion of Geology that it could operate this way. Eventually as the physicists improved their models, Geology was going to become out-manoeuvred, but the Geological Society could only admit this defeat by a capitulation of its traditional authority.

Central to the 1979 discussion on the expanding Earth was the assertion that this theory was no different in form from that of plate tectonics; that the two theories were still vying with each other to win the geologists support. This is a revealing contest. As Holmes first identified, all that plate tectonics requires to be true is that the viscosity and heat production of the Earth is set in certain simple physical limits. In contrast, the expanding Earth requires that matter is increasing in a way that has never been observed and that would upset all known astronomical relationships that have controlled Earth's surface heat through geological time, as well as the physical understandings of matter and space–time. It would be truly revolutionary.

The expanding Earth theory was proposed by a geologist, Warren Carey, and like the early days of Wegener's continental drift, with which it finds great parallels, it is discussed chiefly with reference to Geology and the outlines of maps. That the Moon has shown no expansion validates the theory's geocentricity that satisfies the roots of 19th-century Geology that go back through Lyell to Aristotle: the science of the Earth operates according to laws different from those of the heavens.

The arrival of a new opposition to plate tectonics was more unexpected. In January 1983 the British daily the *Guardian* ran in its science columns[16] a long article claiming that 'the cherished beliefs [of the defenders of plate tectonics] have come under sharp illumination simply from the reflected light of an alternative theory that is mathematically and geophysically more complete'.

The contracting Earth, after decades of neglect had resurfaced, in a book written by the Cambridge emeritus Professor of Theoretical Astronomy, R. A. Lyttleton.[17] The evidence came from observations and contemplations of the stars; contraction was a universal principle. Lyttleton's 'Earth' has been found lying in the dusty attic of the imagination where it had been abandoned by the German geologists of the 1920s: in identical fashion to Stille's Earth, it has contracted in 20 separate epochs, some of them cataclysmic. 'It is high time' the journalist concludes with perverse irony 'that mathematical rigour returned to the heart of geophysics'.

The Cambridge mathematicians have sustained the 19th century, isolated from any actual contact with the Earth through into the 1980s. For more than 100 years they have studied the Earth as if it were a distant star that could not be visited; as if the dank fens that surround their city are typical of an Earth that does not merit closer scrutiny. The dynasty has survived in perfect line: the contracting Earth that Hopkins taught Kelvin, Kelvin taught George Darwin, George Darwin passed on to Jeffreys and now Jeffreys has taught Lyttleton.

In England the continued survival of the institutionalised Geology Empire and 19th-century Earth Physics is something of an anachronism since they have been almost impotent in affecting the actual course of research. To discover what would happen if that traditional Geology had far greater control over the response to the new theories it is necessary to travel to a different scientific culture: that of the Soviet Union.

The Russian rift

Soviet Geology grew apart from its Western cousins. There was little to be inherited from the Tsarist empire: in 1912 there were only 30 Russian geologists. Yet soon after the 1917 Revolution Geology boomed, at the vanguard of a military-style operation directed against the buried rocks with the aim of liberating minerals to feed the agriculture and industry of the Socialist Mother Country. In the desire to turn the Soviet Union into a major industrial power, economically independent of the West, Geology assumed the greatest prominence. The crust was not to be explored but conquered. One representative of the group that styled themselves the 'Young Geologists of Moscow' in 1937[18] described the process as 'a storming of the entrails of the Earth', his own ambition being 'to join the ranks of the fighters on the geological front'. It was a geological hammer that accompanied the sickle on the party flag; but a hammer that was a weapon rather than an instrument of enquiry.

Many were trained to operate in cadres as the simple foot-soldiers in these battles. By 1967 there were 112000 geologists, of whom 64078 had been to universities and colleges. (In the United States at this period there were around 20000 active geologists.) In growing from 30 to more than 100000 this new Red Army of the Earth had of necessity to suppress the personal whims of scientific enquiry in favour of simple national objectives. In the style of the military, the vast majority of geologists were concerned only with locating resources; while speculation and theory building were left to a tiny coterie of 'generals'.

In the early years of the Revolution there had been a number of Soviet geologists who had voiced sympathy for the brash new theory of continental drift, but as the Revolution became institutionalised, so this enthusiasm 'died quickly', and there was a return to the older conservative models. In 1934 the geologist M. M. Tetyayev founded the new Soviet school of Earth theories that was entirely based on the reactionary fixism of the geologists; of Stille and Kober as modified and exaggerated in the works of Haarmann with his denial of all horizontal motions, and Bucher with his pulsating Earth alternating between the diastolic and systolic phases of its clumsy buried heart. Tetyayev wrote a book, *The foundations of Geotectonics*, which in the post-revolutionary vacuum was to have primacy as the one true theory. The moment in time, the culture, defined the structure of this book that was itself to provide much of the source and foundation for future Soviet thinking. In Tetyayev's picture of the continents, crustal blocks went up and down like the keys on a Pianola.

As post-revolutionary Russia entered its late adolescence there was more and more pressure for scientific theories to orientate themselves to theoretical marxist views. At first this involved disavowing foreign ideas that had been nurtured within a capitalist system. As the new society became more ordered so it became necessary to explore all those ramifications of marxism. No longer could science be seen simply as the means of improving general welfare; dialectical material- ism was the 'Absolute Creator', and in a campaign orchestrated by the speeches of Josef Stalin, parallels had to be found between all historical sciences, such as Evolutionary Biology and Geology, and the one true pathway through time: the marxist 'theory of history'. For marxism to be true required this pathway to be fundamental to the very structure of the Universe.

Within Geology the process of ideological refining was not so deleterious as in Biology (where genetics was outlawed); there was no established model to be banned and no simple interconnection between such theories and practical prospecting and mining. There was, however, amidst a spate of rhetoric the denouncement of ideas that were 'formalistic', 'bourgeois', 'metaphysical' and 'idealistic'.

In 1937 the Soviets chose to put themselves on display by hosting the 17th International Geological Congress; a highly successful occasion at which the slow pace of geological advancement was for once enlightened by the excited optimism of this young nation. The Soviet chairman praised his heroic country and the great Stalinist epoch, and the Spanish Republican delegate spoke fervently against the retardation of science when imposed on by imperialism. The new theoretical Soviet Geology made its appearance in a paper delivered by one member of the Soviet organising committee: M. A. Usov.[19] Tetyayev's and Bucher's rhythms had gained an ideological inter- pretation: 'Geotectonics' he stated 'is a manifestation of the process of self-development of the Earth's matter. This process progresses as a result of a struggle between two conflicting factors immanent to the Earth's matter, compression and expansion . . . . (L)ong periods of suppressed struggle . . . are followed by a revolutionary phase.' As with the proletariat, the forces of containment in the Earth were within the upper crust. Usov considered that attempts at geotectonic generalisation outside the Soviet Union 'verged on metaphysics'.

Three years later the President of the Soviet delegation at that congress, Academician W. A. Obruchev, criticised both Bucher's and Usov's models as not according fully with dialectical materialism.[20] Within Obruchev's revision the Earth's struggle was not suppressed; the constant dialectical conflict between the opposing forces was relieved by slow oscillatory movements. Soon after this, and in the midst of the Second World War, Vladimir Vladimirovich Beloussov, a younger geologist and acolyte of Tetyayev, proposed a mechanism for the up–down oscillation, based on the periodic melting originally proposed by Joly. Haarmann's *oszillations-theorie* had presumed unknown cosmic influences as its cause. The explanation could now be brought down to Earth. The migration of colonies of radioactive elements within and below the continents caused changes in heat flow that would contract or expand the crust. After contraction the sediments would infill the hollow, changing the population of radioactive elements and eventually cause the crust to reheat and rise like a soufflé. The oscillations were slow and regular; each modern-day mountain range represented just one cycle captured towards its climax.

Beloussov was a contemporary of the great quartet of geophysicists: Ewing, Bullard, Hess and Wilson. Born in 1907 he had investigated natural gas deposits and mapped in several Soviet mountain ranges. He had delivered a paper at the 1937 Congress and had just moved back to his birthplace to become Professor at Moscow State University, when his theories were first published. Beloussov was not a Communist Party member but a kulak from the old order, a Russian who had little time for the excesses of dialectical materialism in geological theory. He was, however, in common with Usov and Obruchev, a fervent Nationalist and therefore well suited to thrive in the shifting ideological climate of the post-war era. His authority and leadership were recognised in 1948 when he led the Soviet delegation of 130 geologists to the 18th International Geological Congress in London and in the same year published his major work *General geotectonics*. As the USSR entered its second phase of isolationism (only 11, undistinguished, Soviet delegates attended the 1952 International Geological Congress in Algiers), Beloussov revised his book to reappear in 1954 as *Basic problems in geotectonics* (English translation 1962), and in 1953 he was elected to become a corresponding member of the prestigious Soviet Academy of Sciences; an honour that was, perhaps because of his distance from the Party, never to become converted into the full status of Academician.

Beloussov's attitudes as a critic and patriot were forged in the purification of the pre-war years. While in the West the self-confidence of the geological whole-Earth theories of the 1920s had given way to a scepticism and introspection (Bucher had recanted his own pulsating heart-Earth), Beloussov affirmed a new strength and certainty, thus providing perfect continuity to the pre- and post-war cultures; the 1950s and 1960s were to be based on the 1930s. 'Many hypotheses of geotectonics' he complained 'have caused considerable damage to geotectonics giving non-specialists the impression that this is a field in which the most superficially conceived fantasy reigns. The clearest example is Wegener's hypothesis of continental drift.' Beloussov's attacks on continental drift were outraged and repetitive. He termed it 'fantastic and nothing to do with science' and claimed that 'these men [converts to Wegener with whom he had passed some student days] were apparently hypnotised by the boldness of Wegener's ideas and by his brilliant style of writing'. 'Oscillatory movements are the fundamental type of tectonic movement. Where are these movements in Wegener's hypothesis?'

In 1957 Beloussov became the official Soviet representative and vice president of the International Geophysical Year, where he was to work so closely with Tuzo Wilson. In 1960 he was elected president of the International Union of Geodesy and Geophysics; he was already director of the Department of Geodynamics in the Institute of Earth Physics of the USSR Academy of Sciences. When in 1964 the Dutch scientific publishers Elsevier instituted an international journal entitled *Tectonophysics* they naturally turned to Beloussov to become the Soviet editor. Up to 1965 he could rightfully claim to be the king of grand tectonic theories; his leadership unchallenged because no new Western spokesmen had emerged. However, the Russian 'Geotectonics' was founded on the extrapolation of geological observations and therefore Beloussov was looking the other way when the British and American Marine Geophysics programmes began to produce their spectacular results.

The first major opportunity for significant contact between the Soviet and Western geologists and geophysicists came in 1968, at a time when the most important plate tectonic papers of McKenzie and

V. V. Beloussov, 1907–
Russian geologist

Parker, Morgan and Le Pichon had already been published. Conferences had always proved to be the most important places for conversions to be made; participants were vulnerable when removed from the orthodoxies of their normal working environment. Yet the 23rd International Geological Congress turned out not to be the site of cultural exchange but instead of political conflict and the most extraordinary piece of theatre that the geological community had ever known. For 1968, the year of revolution in the Earth Sciences, had brought the four-yearly congress to Prague. It began on 19 August with an opening ceremony conducted by President Svoboda. There were over 4000 participants, 300 of them from the USSR, including Beloussov and others who were to become prominent in the ensuing debates: Kropotkin, Khain, Pejve and Zonenshayn. Among the delegations from the West there were Wilson and Laurence Morley from Canada; Brian Harland from Cambridge; Dietz, Heezen and Morgan from the USA.

On the first day of the conference, 20 August, the first of the speakers in the first of the programmes was Beloussov introducing his own 'Geotectonic problems'. At 11 p.m. that evening Soviet troops landed at Prague airport and joined up with tank battalions from the other Warsaw Pact partners to complete their occupation of the city by the following morning. Many of the geologists were awoken by the sound of gunfire. The hall where the General Assembly meetings had taken place was over-run by the invading armies, and the conference was reduced to a few *ad hoc* meetings in the Technical University, the only major building remaining in Czech hands. At the first emergency meeting, with many of the delegates isolated in their hotels through want of any transport, representatives of 26 countries were asked whether they wished to continue. They voted to persevere, though already the Soviet delegates, some in tears, had been encouraged to melt away; Beloussov's own 'International Upper Mantle Committee' had to meet without him.

By the evening of the 21st the Western delegates had begun to agonise over their own morality,[21] varying their stance from the phlegmatic British delegate Peter Kent, who considered that the session should continue unless it would cause embarrassment to the organising committee, to the French delegate Marcel Roubault who had in 1953 attempted to liberate Geology by founding the journal 'Sciences de la Terre': 'one cannot speak of scientific problems in the presence of guns and tanks'. Some geologists offered their blood to the wounded Czechs. Outside the hall the tanks patrolled menacingly, silencing speakers with their roar. The Congress banners were replaced by black flags of mourning; another black flag hung limply over the doorway.[22] Many Congress members painted black bands across their badges. The Congress survived a few days longer simply through the lack of any means of transport to escape from the city. It was eventually concluded on 23 August after most geologists had already left. The enormous effort of preparation dedicated by the

The International Geological Congress, Prague, 1968: *top*, recess of the council meeting on the first day of the Congress, Beloussov second from the left; *bottom*, delegates voting at an *ad hoc* session for the premature closing of the Congress – note the black mourning stripe on the far right (From Congress Proceedings[21])

Czechs had come to nothing; there were only messages of felicitation and sympathy to be delivered; the French delegate abhorred the 'brutal occupation', the Belgian invoked memories of his own small nation's suffering at the whim of bullying oppressors. Of the Eastern Bloc only Romania, the one country that had not contributed to the occupation, had stayed long enough to thank its neighbour. The Soviet military contribution to the conference ensured that there was no transfer of global tectonic ideas across the rift that had now developed between East and West.

Although the Prague Congress would have provided exchange of the new theories there were to be other points of contact. Wilson had sent a copy of his declaration of 'A revolution in the Earth Sciences' to Beloussov. Beloussov's initial response was shocked, stunned: 'The geology of the continents is simply and completely annihilated. Could it be that you really want to rewrite the textbooks, and throw overboard a great part of the outstanding achievements in the geology of the continents?'[23] In 1970 Beloussov published in *Tectonophysics*, his outlet to the West, a long and vituperative attack against plate theory.[24] His position was summarised in a single sentence: 'The presently fashionable hypothesis of ocean-floor spreading is examined and the conclusion is drawn that this hypothesis is unacceptable.' Yet the diatribe continued 'Not a single aspect . . . can stand up to criticism . . . hasty generalisation . . . monstrously overestimated . . . replete with distortions. . . . It brought to the earth sciences an alien rough schematization permeated by total ignorance of the actual properties of the medium.' Again, all geotectonic phenomena demonstrated *regular* interrelationships.

As a result of the IGY Beloussov had become an international geologist; the only geologist in the Soviet Union who consistently chose to publish in the West. Between 1970 and 1973, in parallel with Western nations, the Soviets organised a large expedition to Iceland. For the Americans this was to see the mid-Atlantic spreading ridge exposed on dry land. The Soviet expedition leader was Beloussov. On completion of his team's work Beloussov, in 1977, wrote 'any drifting apart of the continents fringing the Atlantic Ocean (at the latitude of Iceland) is excluded'.[25] In place of sea-floor spreading that the American expeditions considered they had validated, Beloussov still supported Barrell's 'oceanisation' in which blocks of continental crust have foundered and been overwhelmed by oceanic magmas.

In 1972 Beloussov found himself cornered by the seven-man International Upper Mantle Committee of which he was the chairman. From 1964 the programme had undertaken investigations into three topics: continental margins and island arcs, the world rift systems and laboratory studies of rocks considered typical of the upper mantle. Much against Beloussov's wishes Tuzo Wilson had cheekily inserted a fourth goal: 'to prove whether or not continental drift occurred'. Eight years later the committee was attempting to draft a document that would be worthy of the revolution in thought that had taken place during its period of operation. The first two sentences reveal the tension within the committee; they overflow with enthusiasm and then become suddenly restrained by conditionalities:[26] 'The period of the U.M.P. . . . witnessed an extraordinary development in earth sciences that is widely considered a "revolution": the emergence of a unifying concept of plate tectonics. It should be emphasized that data collected *at the present* testify in favor of plate tectonics, but that the final decision . . .', etc. The document had required subtle editorial amendment by the committee's secretary general, Leon Knopoff, before Beloussov reluctantly agreed to sign.

In the 1975 edition of the *World Soviet Encyclopedia* the balanced but sceptical articles on plate tectonics are shared and, although independently signed, appear to represent the compounded views of Beloussov and the second signatory who is also one of the theory's most influential sympathisers: the geophysicist Peter Nikolaevich Kropotkin. Born in 1910 Kropotkin was a relative of the great anarchist and geographer, Prince Kropotkin. Peter Kropotkin had sustained an interest in continental drift even at its most unfashionable period. Like Beloussov, Kropotkin had spoken at the 1937 Congress and subsequently became a professor at Moscow University. Since 1959 he has headed the Tectonic and Geophysics Laboratory of the Geological Institute of the Academy of Sciences. Unlike Beloussov his reputation is internal to the USSR; he edits the journal *Geotekhnika*.

For Kropotkin, as for many other Western members of his generation, plate tectonics is merely the second wave of continental drift. During the late 1950s there was a very small, but highly successful, Soviet research effort in palaeomagnetism pioneered by Aleksei Khramov who believed both that it was possible to erect a timescale of magnetic reversals and that the polar wandering curves for rocks from

P. N. Kropotkin, 1910–
Russian geophysicist

different parts of the Soviet Union indicated that the continent had once been inchoate. His most impressive result, which was to be echoed by Kropotkin, was that the Siberian and Russian tectonic platforms had been separated by at least 3000 km before the two came together to push up the Urals. The first demonstration of Kropotkin's sympathy for the new theories was elicited in a letter to his friend, the Cambridge crystallographer J. D. Bernal, which was published in the Proceedings of the 1963 Symposium on Continental Drift held by the Royal Society of London. Kropotkin cites three lines of evidence that lend support to drifting continents: the Soviet palaeomagnetic data, the evidence of the earthquakes and, in the style of Wegener, the changes in geographical latitude between Leningrad and western USA made from astronomical observations. There is no suggestion of any future contribution from Marine Geophysics but some optimism that 'the drift of continents represents a problem in which the co-operation of scientists of different countries can be especially fruitful'.

Kropotkin continued to sustain an enthusiasm for the Western discoveries; writing such papers as 'The mechanism of crustal movement'[27] in 1967 and in 1971 'Eurasia as a composite continent'[28] in which the Union of the Soviets was seen to have taken place during geological history through the collision of several smaller fragments of continent. In this year his lonely championing of drift and the new global tectonics won support from another important spokesman, L. P. Zonenshayn from the Science Research Laboratory of Geology of Foreign Countries in Moscow. Zonenshayn, following the example of Dewey, began to interpret all the geosynclines known to have existed in Central Asia as products of sea-floor spreading. In 1973 another high-placed supporter, Academician V. I. Smirnov (Director of the Department of Geology, Geophysics and Geochemistry within the Academy of Sciences) instigated a conference on 'Metallogeny and the new global tectonics' in Leningrad. It remained until 1974 for Russians to learn of the great changes through the translation of a series of the most important Western papers on sea-floor spreading and plate tectonics in a work edited by Zonenshayn entitled *The new global tectonics*.

Still the advocates of mobilism remained very much in the minority. Not until 1977 and 1978 were there signs of a growing interest and a widespread debate. Yet, unlike the situation in the West, the appearance of more geologists prepared to sanction plate tectonics provoked a powerful backlash. For example Ye. M. Rudich's book *The Atlantic Ocean and continental drift*, published in 1977, was devoted solely to disproving all evidence of drift in the South Atlantic. At the end of January 1978 a conference was held in Moscow on the Mediterranean Belt,[29] which provided a forum for wide-ranging opinion. One of the most prestigious of the new global tectonics converts was Academician V. Ye. Khain (Professor at Moscow University, born in 1914, author of *General Geotectonics*, 1964) who detailed views more or less compatible with those of Western Earth scientists. Yet at the same confer-

218

ence, Beloussov sustained his vertical oscillation tectonics; Ye. Ye. Milanovskiy (from the Geological Faculty of Moscow State University), who had written the conclusions to the Icelandic investigations with Beloussov, supported a pulsating globe with episodic compression and tension as well as an overall planetary expansion, and Yu. V. Chudinov believed all could be explained with expansion alone. Kropotkin, his impatience showing through the bland text of the conference report, remarked on the inadequate use of geophysical methods by geologists.

To find the balance of opinion one need only read Soviet self-perception: Milanovskiy writing in 1978[30] about what he terms 'neomobilism' considers that 'this concept ... is shared by *some* Soviet geologists and geophysicists' (my emphasis). Amidst the increasing receptiveness of the Soviets there has been a desire to show either that the new global tectonics is merely the return of continental drift, that the West itself effectively outlawed, or else, more significantly, that plate tectonics is not yet the full story. In 1978 the eminent Academician V. Ye Khain wrote an article entitled 'From plate tectonics to a more general theory of global tectonics'.[31] A Western Earth scientist would argue that plate tectonics was never intended to describe anything other than the broadest scale of Earth behaviour. 'Renaissance of the ideas of mobilism' writes Khain '. . . had a progressive influence, on the whole, on the history of geotectonics and geology, despite a certain rigidity in the early plate tectonics'. This is sophisticated irony, a mobilist theory is termed in common with the lithospherical plates: 'rigid'. Yet this new attack is being delivered by Khain as a shrewd political manoeuvre: 'The hope of further progress in tectonic science rests in overcoming the rigidity and establishing a broader theory organically incorporating some of the fixist concepts.' Khain is anticipating some conciliation within the deep divisions that cut through Soviet geotectonic theory. He ends his article with what reads as an admission of past neglect: 'It goes without saying that Soviet students of tectonics can and must make a tangible contribution to answering these questions.'

The following year Beloussov, still an immovable fixist, has mellowed, refined his rhetoric and educated himself in data on plate tectonics. After listing his arguments in the weekly news-sheet of the American Geophysical Union[32] he concludes 'something will certainly remain after the theory of plate tectonics goes. Let us keep our minds open and look for alternatives. I am sure that in the near future we will need them.'

At the end of August 1979 the international Pacific Science Congress was held at Khabarovsk,[33] involving a total of 1200 Soviet delegates with 500 from other nations ringing the Pacific. Tectonics proved the major topic of the Congress but even in the presence of an international audience the Soviets sustained their nervous eschewing of the, by now, decade-old theories. Corresponding Member of the Academy of Sciences Yu. M. Pushcharovskiy 'noted three trends in modern tectonics: rigid fixism, plate tectonics and the trend of developing classical mobilism.' The group of mobilists at the Congress's first symposium 'was less numerous' than 'the several reports [that] clearly showed a fixist tendency'. In the second symposium on the Pacific Ocean itself, Khain, Zonenshayn and others who had worked on shipboard studies supported sea-floor spreading explanations; but in the ensuing discussion 'many of the participants drew attention to facts and situations which contradict the hypothesis'. It was left to the minority of converts to attempt to battle this hostile reception; Zonenshayn in some annoyance 'exhorted their opponents to study the fundamentals of the hypothesis and to argue from a knowledge of the facts.'

In the write-up given to the 1980 26th International Geological Congress in Paris,[34] Kropotkin, in his own *Geotektonikha* describes how 'in an even greater degree than previous sessions', the Congress took place 'in an atmosphere of almost complete acceptance of mobilism'. He lists the exceptions: Beloussov is now in the company of only one other geologist, Chan Venyu, the Director of the Geological Institute of the Chinese Academy of Sciences. This is, however, a very partial view. Both he and Beloussov had chosen to present their contributions in the 'Geophysics' sessions. Within 'Tectonics' the true extent of the decline in Soviet authority is apparent. Out of 424 papers presented, they offer only 11 among which there are twice as many proposing some alternative to plate tectonics in place of welcoming it. The quasi-official Soviet attitude of scepticism was represented by the new-found power of Yu. M. Puscharovskiy of the Geological Institute of the Academy of Sciences who talks of 'the great anciety of the Pacific Ocean' and prefers some process whereby continental crust can become oceanised. In the Special Colloquium on the 'Geology of the oceans' there were no Soviet contributions.

In 1982 in a Moscow radio broadcast[35] it was admitted still that 'many Soviet geologists do not support plate tectonics'. How had the Soviet Union arrived at this point, enjoying as it could claim a degree of pluralism about tectonic theories that was denied in the regime of plate tectonic authoritarianism of the West? In 1965 the composition of intellectual attitudes to crustal mobilism showed an identical prejudice in the USSR to that pertaining in the USA. Supporters of drift had been termed 'leftist' in American and 'bourgeois' in the Soviet Union. To the West there was a Hess, to the East a Kropotkin. Both great continental nations had throughout the history of their geological investigations concentrated on the geology to be found within their frontiers and thus had equally tended to stress the importance of stable continental platforms and vertical movements of the crust. In both the Soviet Union and the USA theories had often been blessed by a nervous nationalism as 'American' or 'Russian', and therefore superior to outside pretenders. In both countries the role of the geologist in pioneering the exploration of the virgin interior had given to Geology a special privileged self-importance that was missing from civilised and settled Europe. In both countries the economy had been built on the resources pulled out of the ground. However, as the development of the mineral resources of America had begun a full half-century before those of Russia so the roles of the geologist were out of phase. Chamberlin could only have gained his power base in the first rapid growth of the Science at the vanguard of the exploration and development of the frontier lands. Beloussov is Chamberlin's equivalent, gaining his power and self-certainty through the same period of the Science's growth. The similarities between these two extend to the very grammar of hostility that Beloussov and Chamberlin's proxy, his son Rollin, use to condemn Wegener. 'Wegener's theory, which is easily grasped by the layman because of its simple conceptions, has spread in a surprising fashion among certain groups of the geological profession' complained Chamberlin. 'Wegener's hypothesis . . . has caused considerable damage to geotectonics giving non-specialists the impression that this is a field in which the most superficially conceived fantasy reigns' grumbled Beloussov. If there had been a Chamberlin within Geology in America in the 1960s then plate tectonics would not have been allowed to overturn the older science without a terrible struggle. But Geology in America had become dissipated and weak.

Yet what has happening outside Geology; why had not some formative Soviet Earth Sciences attempted to mount a palace coup? How had America and Russia come to drift so far apart? The revolution in America came from the wealth of new post-war research topics, and was triggered by the discoveries emerging from ocean Geophysics. Yet during the 1950s and 1960s the Soviet Navy, with a less defined strategic role, never gained the massive scientific research programme to assist in its global operations. The very size of the country has influenced the perspective: the Soviet Union occupies one-sixth of the land area of the Earth. Sea-floor spreading has few implications for a land-locked Muscovite. Soviet oceanographic research had not concentrated on mapping the geophysics of the oceans, with one exception: the Arctic.

From 1961 to 1966 A. M. Karasik and R. M. Demenitskaya, of the Institute of the Geology of the Arctic in Leningrad, performed a variety of geophysical and bathymetric investigations on the mid-Arctic ocean ridge which they found was the axis to an extensive pattern of magnetic anomalies. From 1966 to 1968 they had included the pattern of seismicity obtained at Lamont in with their data, and in collaboration with a number of other researchers produced a series of papers that were almost identical to those emerging from marine geophysicists in America.[36] However, this was a remarkable exception to a prevailing resistance to change, and these papers on sea-floor spreading from the Leningrad Institute seem to have dried up soon after 1969, without the second stage conversion to plate tectonics.

In 1973 in the thaw of *détente*, under the diplomacy of Nixon, the USSR agreed through the Institute of Oceanography of the Academy of Sciences to become involved with, and part finance, the Joint Oceanographic Institutions for Deep Earth Sampling; the drilling project from *Glomar Challenger* that was busily confirming sea-floor spreading. The Soviet scientists involved in the cruises of this programme were thrown into the company of French, British and American researchers who all shared a belief and enthusiasm for plate tectonics. As was demonstrated by the Soviet proposals made for future drilling sites,[37] the conversions seem complete. Other Soviet oceanographic vessels were also investigating the deep-ocean floor, carrying scientists who did not always support plate tectonics, sometimes preferring, as A. V. Il'in in 1978,[38] to sustain the view that 'the oceanic relief is shown to be inconsistent with the chief postulates of . . . the new global tectonics'. In 1982 the USSR pulled out of future international collaboration on deep-sea drilling.

Throughout the Western world there has been some kind of revolution, more acute in America, more subdued in Europe. Yet in the Soviet Union, there was no such transformation. What illumination does this throw upon the intellectual structure of Soviet science? To understand, it is necessary first to find the Soviet perspective. Western Earth scientists appeared to have embraced the new-born doctrine in the way they adopted all the fashions and tinsel of capitalism: with wild irrational enthusiasm. Plate tectonics was the most glamorous thing to have happened to Geology since the Flood. To have climbed on the bandwagon early offered young researchers a short cut to scientific glory. Not to mention plate tectonics in a geological paper showed a wilful neglect of novelty. As Beloussov wrote in an open letter to Wilson in 1968, 'It would be irresponsible of us to tempt young people saying that all the difficulties are behind us and, instead of leading them along a hard and strenuous path of search and menial labour, inevitable for a scientist, lull them with delusive hopes and dreams.'[23]

The 'hard and strenuous path of search and menial labour' trains the prospectors, the working geologists; ruled over by the geotectonicians. To undermine that hierarchy would be to undermine a section of the gerontocracy. It seems to have been impossible for the young research student, the Vine, Morgan or McKenzie to reshape a science. To replace the Soviet hierarchy of Geology will require gradual demolition and reconstruction over many years; an opposition to plate tectonics will probably be sustained into the 21st century. For the hierarchy of geological authority is facing a scientific revolution that demands a complete reorganisation in the power structure whereby the old geologist-geotectonicians give way to the new global-geophysicists, thus creating Earth Sciences. The geological hierarchy cannot be deposed or out-manouvered because it is as rock-solid as a continent. The immovable object of Soviet bureaucracy meets the irresistible force of a scientific revolution. The contest is titanic – yet even continents may drift. Thus the history of Soviet attempts to accommodate plate tectonics has implications going far beyond some simple story of scientific controversy. It provides a visible model of how that greater bureaucracy, the State, can hope to adapt itself to the need for radical transformation.

Yet through constructing an elaborate and massive geotectonic cosmology, both numerically and critically, Soviet Geology remains in excellent health, preserving a strength and dignity of Geology that would not shame the great age of the mid 19th century. Research rides high on unmanned missions to Mars and Venus. Soviet geologists have been enabled to broaden their scope into planetology more readily than in the West, where the new breed of planetologists are likely to turn up in astronomy departments. The deep structure of the American continent is being mapped by geophysicists while in the USSR the geologists have the political power to gain support for an immensely costly borehole more than 12 km into the crust. Thus the old empire of Geology in danger of extinction in the West can hope to survive, untrammelled.

G-éloge

In the West the desire for change overwhelms any evaluation of the value of previous knowledge and research. What Earth scientist would wish to spend years compiling an accurate and detailed map of one small region? Yet just as Natural History continued to survive after Darwin, so parts of the old Geology no longer considered suitably scientific will be sustained through the amateur. The creation of the Earth Sciences has broken the pattern that allowed Geology to survive informal and unhurried for more than a century. With the loss of the older science there will also go the expert whose knowledge consists of the ability to name a plethora of fossils, minerals, and identify the stratal age; all these skills are denigrated in the new overviews of the Earth scientist involved in laboratory work or indirect geophysical mapping. Thus as the science gains a greater appreciation of the whole planet so it loses the joyous intimacy with the landscape that came from the purposelessness of summer fieldwork and winter walks along the seashore, an intimacy that lies at the very heart of Hutton's philosophy of Geology. It is the passing of an era: through becoming scientific, the old Geology, that was so framed according to the poetic and irrational measure of man, is now, in the new technological Earth Sciences, in danger of losing its soul.

References

Chapter 1

1   Playfair, J. 1811. Review of *Trans Geol Soc. London. Edin. Rev.* **xix**, 207–29.
2   Weld, C. R. 1848 (quoting Whewell, W.). *A history of the Royal Society*, 245. J. W. Parker, London.
3   Phillips, W. 1815. *An outline of geology and mineralogy*, Collins, London.
4   Paris, J. A. 1818. Observations on the geological structure of Cornwall. *Trans R. Geol Soc. Cornwall* **i**, 168–200.
5   'Geology' in *Rees Cyclopaedia* (1819).
6   Darwin, C. 1933. *Diary of the Voyage of H.M.S. Beagle* N. Barlow, ed. Cambridge University Press, Cambridge.
7   Anon. 1982. Obituary notice of D. G. Jones. *Proc. Geol Assoc.* **92**, 210
8   Hall, J. 1859. Natural History of New York. *Palaeontology* **3** (1), Albany, New York.
9   Dana, J. D. 1875. On the origin of mountains. *Am. J. Sci. 3rd Series* **5**, 347–50.
10 Gilbert, G. K. 1877. Report on the geology of the Henry Mountains, Government Printing Office, Washington DC.
11 Haughton, S. H. 1948. Alexander Du Toit 1878–1948. *Biog. Mem. Fellows R. Soc. Lon.* **6**, 385–95.
12 Schmidt, H. 1973. As quoted in *Continental drift – the evolution of a concept*, by U. B. Marvin, preface. Smithsonian Institution Press, Washington DC.
13 Hoyle, F. 1950. Nature of the Universe; a series of broadcast lectures. B. Blackwell, Oxford.

Chapter 2

1   Elie de Beaumont, L. 1866. *Carte géologique générale de la France*. Impr. impériale, Paris.
2   Cuvier, G. L. C. F. D. 1825. Discourse sur les révolutions de la surface du globe, Dufouret d'Ocagne, Paris.
3   Elie de Beaumont, L. 1829. Recherches sur quelques unes des révolutions de la surface du globe. *Ann. Sci. Nat.*, **18**, 5–25.
4   Newton, Sir Isaac 1681. Letter to Thomas Burnet, Master of Charterhouse. In *The Correspondence of Isaac Newton*, vol 2, 329–34. Cambridge 1960. Ed. H. W. Turnbull. Cambridge University Press, Cambridge.
5   Elie de Beaumont, L. 1842. Note inserted in translation of 2nd edn. *Manuel géologique* by Henry de la Beche. Langlois et Leclercq, Paris.
6   Elie de Beaumont, L. 1850. Note sur la corrélation des directions des différent systèmes de montagnes. *C. R. Acad. Sci. Paris* **31**, 325–38.

References

7 Green, W. L. 1857. The cause of the pyramidal form of the outline of the southern extremities of the great continents. *Edin. New Phil J.* **6**, 61–78.

8 Darwin, C. R. 1838. On the connection of certain volcanic phenomenon in South America. *Trans Geol Soc. Lond. 2nd Series* **5**, 601–632.

9 Mallet, R. 1873. Volcanic energy: an attempt to show its true origin and cosmical relation. *Phil Trans R. Soc. Lond.* **163**, 147–227.

10 Prinz, W. 1891. Sur les similitudes que présentent les cartes terrestre et planétaires. *Ann. l'Obs. R. Brux.* **58**, 304–37.

11 Gregory, J. W. 1899. The plan of the Earth and its causes. *Geogr. J.* **XIII**, 225–51.

12 Hitchcock, C. H. 1900. William Lowthian Green and his theory of the evolution of the Earth's features. *Am. Geol.* **XXV**, 1–10.

13 Hopkins, W. 1838. On the state of the interior of the Earth. First and second memoirs. *Phil Trans R. Soc. Lond. Abs.* **4**, 83–4, 115.

14 Thomson, W. 1861. Physical considerations regarding the possible age of the Sun's heat. *Brit. Assoc. Rep.* 27–8.

15 Fisher, Rev. O. 1873. On the formation of mountains with a critique of Captain Hutton's lecture. *Geol Mag.* **X**, 248–61.

16 Fisher, Rev. O. 1874. On the formation of mountains viewed in connection with the secular cooling of the Earth. *Geol Mag. New Series* **1**, 60–3.

17 Fisher, Rev. O. 1909. On convection currents in the Earth's interior. *Geol Mag.* **6**, 8–11.

18 Chamberlin, T. C. 1904. The methods of the Earth Sciences. *Pop. Sci. Monthly* **66**, 73.

19 Chamberlin, T. C. 1899. Lord Kelvin's address on the age of the Earth as an abode fitted for life. *Science* **9**, 889–901; **10**, 11–18.

20 Pyne, S. J. 1980. *Grove Karl Gilbert, a great engine of research*, 198. Austin: University of Texas Press.

21 Chamberlin, T. C. 1904. Fundamental problems of geology. In *Year Book 3*, 753–5. Carnegie Institution. Washington DC.

22 Chamberlin, T. C. 1909. Diastrophism as the ultimate basis of correlation. *J. Geol.* **17**, 685–93.

23 Willis, B. 1929. Memorial of Thomas Chrowder Chamberlin. *Bull. Geol Soc. Am.* **40**, 23–30.

24 Suess, E. 1885. *Das Antlitz der Erde*. Translated in English, by H. C. B. Sollas, 4 vols. Oxford: Clarendon Press. **1**, 604.

Chapter 3

1 Dana, J. D. 1866. Observations on the origin of some of the Earth's features. *Am. J. Sci.* **42**, 205–11.

2 Humboldt, A. von. 1801. Esquisse d'un tableau géologique de l'Amerique méridionale. *J. Phys. Chem. d'Hist. Nat.* **53**, 30–60.

3 Bacon, F. 1620. *Novum Organum*, Book 2, Aphorism 27. J. Billium, London.

4 Milius, A. 1667. *De origine animalium et migratione populorum*. P. Columesium, Geneva.

5 Hornius, G. 1669. *De originibus Americanis*. J. Müllerli, Hemipoli.

6 Kircher, A. 1675. *Arca Noë, in tres libros digesta. fol.* J. Jansonium, Amstelodami.

7 Hale, Sir M. 1677. *The primitive origination of mankind*. W. Shrowsbery, London.

8 Allen, D. C. 1949. The legend of Noah, *Univ. of Illinois Press* **XXXIII** (3–4). Illinois Studies in Lang. and Lit. **33** iii, iv. Urbana, Illinois.

9 Placet, F. 1666. *La Corruption du grand et petit Monde, section IV*. Paris: Alliot.

10 Lilienthal, Th. Ch. 1756. *Die gute Sache . . . der göttlichen Offenbarung VII*. Königsberg: J. H. Hartung.

11 Franklin, B. 1782. Letter to Abbé Soulavie, 22 Sept. 1782. *The writings of Benjamin Franklin* A. H. Smyth, ed. 1906. New York: Macmillan.

12 Marchant, J. 1916. Letter to Henry Bates. In *Alfred Russel Wallace: Letters and reminiscences*. 2 vols., Harper & Bros., London.

References

13 Sclater, P. L. 1858. On the general distribution of the members of the class Aves. *J. Linnean Soc.* **2**, 130–45.

14 Wallace, A. R. 1859. Letter from Mr Wallace concerning the geographical distribution of birds. *Ibis* **1**, 449.

15 Sclater, P. L. 1864. The mammals of Madagascar. *Q. J. Sci. Lond.* **I**, 213–19.

16 Neumayr, M. 1885. *Erdgeschichte vol 2., Beschriebende Geologie* 2nd ed. Leipzig and Wien: Bibliographisches Inst.

17 Ihering, H. von. 1907. *Archhelenis und Archinotis, Gesammelte Beiträge zur Geschichte der Neotropischen Region*, W. Englemann, Leipzig.

18 Willis, B. 1932. Isthmian links. *Bull. Geol Soc. Am.* **43**, 917–52.

19 Pratt, J. H. 1855. On the attraction of the Himalayan mountains. *Phil Trans R. Soc.* **145**, 53–100.

20 Airy, G. B. 1855. On the computation of the effect of attraction. *Phil Trans R. Soc.* **145**, 101–4.

21 Pratt, J. H. 1859. On the deflection of the plumb-line in India. *Phil Trans R. Soc. Lond.* **149**, 745–78.

22 Dutton, C. E. 1881. The physical geology of the Grand Cañon District. *US Geol Survey Second Ann. Rep.*, 47–166, Washington.

23 Dutton, C. E. 1883. Review of Fisher's *Physics of the Earth's crust*. *Am. J. Sci.* **123**, 283–90.

24 Dutton, C. E. 1887. The Charleston earthquake of August 31, 1886. *US Geol Survey 9th Ann. Rep.*, 1887–8, 203–528.

25 Dutton, C. E. 1889. On some of the greater problems of physical geology. *Bull. Phil Soc. Washington* **11**, 51–64.

26 Chamberlin, T. C. 1888. The rock-scorings of the great ice-invasion. *US Geol Survey 7th Ann. Rept*, 1885–6, 147–248.

27 Gilbert, G. K. 1890. *Lake Bonneville*. US Geol Survey Monographs. Washington, DC.

28 Davis, W. M. 1926. Biographical Memoirs of Grove Karl Gilbert (1843–1918). *Nat. Acad. Sci. Bibliog. Mem.* **21**, No. 2. Washington, DC.

29 Pyne, S. J. 1982. *Grove Karl Gilbert*. University of Texas Press, Austin, Texas.

30 Taylor, F. B. 1898. *An endogenous planetary system: a study in Astronomy*. Archer Printing Co., Fort Wayne, Indiana.

31 Taylor, F. B. 1903. *The planetary system; a study of its structure*. Published by the Author, Fort Wayne, Indiana.

32 Escher, van der Linth A. 1841. Über der Geologie des Kantons Glarus und seiner Umgebung, II. Geologische und mineralogische Section. *Versammlung zu Zurich – Verh. Schweiz. naturf. Ges., 26te Vers.*, 53–79.

33 Heim, A. 1878. Untersuchungen über den Mechanismus der Gebirgsbildung. In *Anschluss an die geologische Monographie der Tödi-Windgällen-Gruppe*; Bd **1** & Bd **2**. Basel: Benno Schwabe.

34 Bertrand, M.– A. 1884. Rapporte de structure des Alpes de Glarus de Basin houiller du Nord. *Bull. Soc. géol. Fr.* **12**, 318–30.

35 Törnebohm, A. E. 1883. Om Dalsformatiönens geologiska alder. *Geol. Förh. Stockh. För* **6**, 622–61.

36 Lapworth, C. 1883. On the secret of the Highlands. *Geol Mag.* **10** (New Series 2), 120–8, 193–9, 337–44.

37 Suess, E. 1891. Die Brüche des östlichen Afrika. *Denkschr. k. Akad. Wiss. math-nat. Cl.* **63**, 555–84.

38 Taylor, F. B. 1910. Bearing of the Tertiary mountain belts in the origin of the Earth's plan. *Bull. Geol Soc. Am.* **21**, 179–226.

39 Peary, R. 1910. *The North Pole*. New York: Hodder and Stoughton.

40 Darwin, G. 1879. The precession of a viscous spheroid and the remote history of the Earth. *Phil Trans R. Soc. Lond.* **170**, 447–538.

41 Fisher, Rev. O. 1882. On the physical causes of the ocean basins. *Nature* **25**, 243–4.

42 Pickering W. H. 1907. The place of the origin of the Moon – the volcanic problem. *J. Geol.* **15**, 23–38.

43 Gilbert, G. K. 1892. *Continental problems of Geology*. Smithsonian Report.

44 Willis, B. 1891–2. Mechanics of Appalachian Structure. *Ann. Rep. of US Geol. Survey* **13**, 211–81.

References

45   Mantovani, M. R. 1889. Les fractures de l'écorce terrestre et la théorie de Laplace. *Bull. Soc. Arts Sci. Réunion* 41–53.
46   Wegener, A. 1912. Die Enstehung der Kontinente. *Geol Rundschau* **3**, 276–92.
47   Wegener, A. 1912. Die Enstehung der Kontinente. *Petermanns Geog. Mitteilungen* **58**, 185–95, 253–6, 305–9.
48   Strahan, A. 1912. Obituary of W. H. Pickering, AGM Geol Soc.

Chapter 4

1   Wegener, A. 1905. Die Alphonsinischen Tafeln. Dissertation, Berlin.
2   Wegener, A. 1906. Über die Entwicklung der kosmischen Vorstellung *in der Philosophie. Phil. Math. Naturwissen. Blätter* **4**, 4–5.
3   Wegener, E. 1960. In *Alfred Wegener*, publ. F. A. Brockhaus, Wiesbaden.
4   Wegener, A. 1905. Blitzschlag in einen Drachenaufstieg am Königlicher Aeronautischen Observatorium, Lindenberg. *Das Wetter* **22**, 165–7.
5   Benndorf, H. 1931. Alfred Wegener. *Beiträge Geophys.* **31**, 337–77.
6   Curie, P. and A. Laborde. 1903. Sur la chaleur dégagée spontanément par les sels de radium. *C. R. Acad. Sci. Paris* **136**, 673–5.
7   Joly, J. 1903. Radium and the Sun's heat. *Nature* **68**, 572.
8   Elster, J. P. L. J. and H. F. K. Geitel 1904. Über die radioaktiv Substanz deren Emanation in der Bodenluft und der Atmosphäre enhalten est. *Phys. Zeit.* **5**, 11–20.
9   Adams, F. D. and J. T. Nicolson 1900–1. An experimental investigation into the flow of marble. *Phil Trans R. Soc. Lond.* **195**, 365–401.
10   Oldham, R. D. 1906. The constitution of the interior of the Earth as revealed by earthquakes. *Phil Mag.* **12**, 165–6.
11   Mohorovičić, A. 1909. Das Beben vom 8 1909. *Jahrbuch Meteorol. Obs. Zagreb (Agram) für das Jahr 1909* **9**(4), 1–63.
12   Fisher, Rev. O. 1909. On convection currents in the Earth's interior. *Geol Mag.* **6**, 8–11.
13   Lawson, A. C. ed. 1908 and 1910. *The California Earthquake of April 18, 1906*, 2 vols. Carnegie Institution, Washington.
14   Davison, C. 1936. *Great earthquakes.* Cambridge: Cambridge University Press.
15   Diener, C. 1915. Die Grossformen der Erdoberfläche. *Mitt. k. k. Gessell. Wien* **58**, 239–49.
16   Jaworski, E. 1922. Die A. Wegenersche Hypothese der kontinental – verschiebungen. *Geol. Rundschau* **13**, 273–96.
17   Arldt, T. 1917. *Handbuch der Paläographie 1* Leipzig: Paläktologie.
18   Koch, J. P. and A. Wegener 1919. Nordgrönlands Trift nach Westen. *Astr. Nachr.* **208**, 270–6.
19   Kossmat, F., A. Penck, W. Penck, W. Schweydar and A. Wegener 1921. Die Theorie der kontinental – verschiebungen. *Ziet. Ges. Erdkunde Berlin*, 89–143.
20   Argand, E. 1916. Sur l'arc des Alpes Occidentales. *Eclogae Geol Helvetiae* **14**, 145–91.
21   Argand, E. 192. La Tectonique de l'asie. *C. R. 13e Congrès Géol. Int. Belg.*, 171–372. H. Vaillant-Carmanne, Liege.
22   Gagnebin, E. 1922. Le dérive des continents selon la théories d'Alfred Wegener. *Rev. Gén. Sci.* **33**, 293–304.
23   Bourcart, J. 1924. Les origines de l'hypothèse de la dérive des continents. *Rev. Sci.* **62**, 563–4.
24   Termier, P. 1925. The drifting of continents. *Annual Report* for Smithsonian Institute for 1924.
25   Bragg, W. L. 1968. Trinity College, Cambridge & Manchester University. In *Sidney Chapman Eighty, from his friends*, Akasofu S. I., B. Fogle and B. Haurwitz (eds), 50. Univ. of Alaska, Univ. of Colorado and Univ. Corporation for Atmospheric Research, Boulder, Colorado.
26   Wright, W. B. and F. E. Weiss 1922. *Proc. Man. Lit. Phil. Soc.*, **66**, 20–22.
27   Anon. (W. L. Bragg ?) 1922. Wegener's displacement theory. *Nature* **109**, 202–3.

28 Wegener, A. L. 1922. The origin of the continents and oceans. *Discovery* April, 114–18.
29 Reade, T. M. 1906. Radium and the radial shrinkage of the Earth. *Geol Mag. Ser.* **5**, vol. 3, 79–80.
30 Eddington, A. S. 1922. The borderland of Astronomy and Geology. *Nature* **111**, 18–21.
31 Lake, P. 1922. Wegener's displacement theory. *Geol Mag.* **59**, 338–46.
32 Wright, W. B. 1923. The Wegener hypothesis. *Nature* **111**, 30–1.
33 Cole, G. A. J. 1922. Wegener's drifting continents. *Nature* **110**, 798–801.
34 Lake, P. 1923. Wegener's hypothesis of continental drift. *Geogr. J.* **61**, 179–94.
35 Douglas, G. V. and A. V. Douglas 1923. Note on the interpretation of the Wegener frequency curve. *Geol Mag.* **60**, 108–11.
36 Wright, W. B. 1923. Letter 'The Wegener frequency curve'. *Geol Mag.* **60**, 239–40.
37 Jeffreys, H. 1923. Letter 'Hypothesis of continental drift'. *Nature* **111**, 495–6.
38 Gregory, J. W. 1925. 'Continental drift' (a review of the English translation of the 3rd edn of Wegener's book). *Nature* **115**, 255–7.
39 Coleman, A. P. 1925. Letter 'Permo-carboniferous glaciation and the Wegener hypothesis'. *Nature* **115**, 602.
40 Meyrick, E. 1925. Wegener's hypothesis and the distribution of Micro-lepidoptera. *Nature* **115**, 834–5.
41 Lake, P. 1926. The origin of the continents and oceans: review of English ed *Geol Mag.* **63**, 181–2.
42 Evans, J. W. 1925. Anniversary address of the President. *Q. J. Geol Soc. Lond.* **lxxxi**, cxxii.
43 Gregory, J. W. 1929. The geologic history of the Atlantic Ocean (pres. address). *Q. J. Geol Soc. Lond.* **85**, xviii–cxxi.
44 Green, J. F. N. 1935. Meeting of Jan. 23rd. *Q. J. Geol Soc. London.* **xci**, v–xi.
45 Taylor, F. B. 1920. Some points in the mechanism of arcuate and lobate mountain structures. An objection to the contraction hypothesis as accounting for mountains. *Abs. Bull. Geol Soc. Am.* **32**, 31–4.
46 Pickering, W. H. 1924. The separation of the continents by fission. *Geol Mag.* **62**, 31–4.
47 Reid, H. F. 1922. Drift of the Earth's crust and displacement of the Pole. *Geog. Rev.* **12**, 672–4.
48 Ann Arbor Meeting 1922. Sites and nature of North American geosynclines, C. Schuchert. Outlines of Appalachian structure, A. Keith. *Bull. Geol Soc. Am.* **34**, 151–230, 309–80.
49 Lambert, W. D. 1923. Mechanics of the Taylor–Wegener hypothesis of continental migration. *J. Wash. Acad. Sci.* **13**, 448–50.
50 Melton, F. A. 1925. The origin of the continents and oceans. Review. *Science* **62**, 14–15.
51 Ingalls, A. G. 1926. Are the continents drifting? *Sci. Am.* January, 8.
52 van Waterschoot van der Gracht and B. Willis, R. T. Chamberlin, J. Joly, G. A. F. Molengraff, J. W. Gregory, A. Wegener, C. Schuchert, C. R. Longwell, F. B. Taylor, W. Bowie, D. White, J. T. Singewald Jr., and E. W. Berry 1928. Theory of continental drift – a symposium on the origin and movement of land-masses, both inter-continental and intra-continental, as proposed by Alfred Wegener; held 15 Nov. 1926, by the American Association of Petroleum Geologists at Tulsa.
53 Georgi, J. 1962. Memories of Alfred Wegener. In *Continental Drift*, K. Runcorn (ed.), 309–24. New York: Academic Press.
54 Georgi, J. 1935. *Mid-ice: the story of the Wegener expedition to Greenland.* New York: Dutton.

Chapter 5

1 Joly, J. 1899. *An estimate of the geological age of the Earth.* Smithsonian Reports, 247–88.
2 Joly, J. 1903. Letter 'Radium and the geological age of the Earth.' *Nature* **68**, 526.

References

3 Rutherford, E. 1905. Letter to Boltwood. In *Rutherford and Boltwood: letters on radioactivity*, Badash, L. (1969), 100–5. New Haven: Yale University.

4 Elster, J. P. L. J. and H. F. K. Geitel 1904. Über dei radioaktiv Substanz dern Emanation in der Bodenluft und der Atmosphäre enhalten est. *Physik. Zeit.* **5**, 11–20.

5 Sollas, W. J. 1905. *The age of the Earth and other geological studies.* London: Unwin.

6 Holmes, A. 1913. *The age of the Earth.* New York: Harper.

7 Harker, A. 1915. Geology in reaction to the exact sciences, with an excursus on geological time. *Nature* **95**, 105–9.

8 Holland, T. H. 1914. Presidential Address Section C. *British Assoc. Rep. (Australia),* 344–58.

9 Barrell, J. 1914 and 1915. The strength of the Earth's crust. *J. Geol.* **22**, 28–48, 145–65, 209–36, 289–314, 441–68, 537–55, 655–83, 729–41; **23**, 27–44.

10 Holmes, A. 1931. Radioactivity and earth movements. *Trans Geol Soc. Glasgow* **18**, 559–606.

11 Schwinner, R. 1920. Vulkanismus und Gebirgsbildung. *Zeit. Vulkanologie* **5**, 176–230.

12 Tyrrell, G. W. 1925. Are the continents adrift? *Sci. Prog.* **20**, 321–5.

13 Tyrrell, G. W. 1925. Letter. *Sci. Prog.* **20**, 512.

14 Huxley, T. H. 1869. Geological reform. *Q. J. Geol. Soc. Lond.* **25**, xxxviii–liii.

15 Ampferer, O. 1906. Über das Bewegungsbild von Faltengebirgen. *K. K. Geol. Reischsanst (Vienna) Jahrb.* **56**, 539–622.

16 Stille, H. 1913. *Tektonische Evolutionen und Revolutionen in der Erdrinde.* Leipzig: Veit.

17 Stille, H. 1936. Wege und Ergebnisse der geologisk – tektonischen Forschung. *Festschr. Kaiser-Wilhelm Gesell. Förd. Wiss.* **2**, 77–97.

18 Haarmann, E. 1930. *Die Oszillations-theorie.* Stuttgart: Ferdinand Enke.

19 Lindemann, B. 1927. *Kettengebirge. Kontinentale Zerspaltung und Erdexpansion.* Jena: G. Fischer.

20 Hilgenbirg, O. C. 1933. *Vom wachsenden Erdball.* Berlin: Giessmann & Bartsch.

21 Hilgenbirg, O. C. 1967. 'Why earth expansion?' Talk given at NATO Advanced Study Symposium, Newcastle, 1967 (privately published).

22 Nagl, M. 1973–4. SF occult sciences and Nazi myths. *Sci. Fict. Stud.* **1**, 185–97.

23 Fauth, Ph. (ed.) 1926. *Hoerbiger's glazial kosmogonie*, 2nd edn. Kaiserslauten, H. Kayser: Leipzig.

24 Gutenberg, B. 1927. Die Veränderung der Erdkruste durch Fliessbewegungen der Kontinentalscholle. *Gerlands Beitr. Geophysik* **16**, 281–91.

25 Taylor, F. B. 1932. Wegener's theory of continental drifting; a critique of his views. *Abs. Bull. Geol Soc. Am.* **43**, 173.

26 Willis, B. 1928. Growing mountains. *Sci. Am.* April. 151–4.

27 Baker, H. B. 1911. The origin of the Moon. *Detroit Free Press,* April 23rd. (See also The origin of continental forms, *Mich. Acad. Sci. Ann. Rep.* 1912, 116–41; 1913, 107–13; 1914, 26–33, 99–103.)

28 Mayr, E. (ed.) 1952. The problem of land connections across the South Atlantic, with special reference to the Mesozoic (symposium proceedings). *Bull. Am. Mus. Nat. Hist.* **99**, 79–258.

29 Du Toit, A. L. 1920. Land connections between the great continents and South Africa in the past. *South African Assoc. Adv. Sci.* 120–40.

30 Du Toit, A. L. 1927. *A geological comparison of South America with South Africa.* Carnegie Institute of Washington Publication, No. 381.

31 Daly, R. 1950. Memorial to Alexander Logie Du Toit. *Proc. Geol. Soc. Am. Ann. Rep.* 1949, 141–9.

32 Simpson, G. G. 1943. Mammals and the nature of the continents. *Am. J. Sci.* **241**, 1–31.

33 Joléaud, L. 1919. Relations entre les migrations du genre *Hipparion. C.R. Acad. Sci. Paris* **68**, 177.

34 Du Toit, A. L. 1944. Tertiary mammals and continental drift. *Am. J. Sci.* **242**, 145–63.

35 Longwell, C. 1944. Some thoughts on the evidence for continental drift. Further discussion of continental drift. *Am. J. Sci.* **242**, 218–31, 514–5.

36 Willis, B. 1944. Continental drift, ein Märchen. *Am. J. Sci.* **242**, 509–13.

References

37   Umbgrove, J. H. F., R. D. O. Good, H. E. Hinton, H. Jeffreys, S. W. Wooldridge and J. R. F. Joyce 1951-2. The theory of continental drift. *Advancement Sci.* **8**, 67–88.
38   van Waterschoot van der Gracht and B. Willis, R. T. Chamberlin, J. Joly, G. A. F. Molengraff, J. W. Gregory, A. Wegener, L. Schuchert, C. R. Longwell, F. B. Taylor, W. Bowie, D. White, J. T. Singewald Jr., and E. W. Berry 1928. Theory of continental drift – a symposium on the origin and movement of land-masses, both intercontinental and intra-continental, as proposed by Alfred Wegener; held 15 Nov. 1926, by the American Association of Petroleum Geologists in Tulsa.
39   Mercanton, P. L. 1926. Inversion de l'inclinaison magnetique terrestre aux äges géologiques. *Terr. Magn. Atmos. Elec.* **31**, 187–90.
40   Elsasser, W. 1946. Induction effects in terrestrial magnetism. *Phys. Rev.* **69**, 106.
41   Runcorn, S. K., A. C. Benson, A. F. Moore and D. H. Griffiths 1951. Measurement of the variation with depth of the marine geomagnetic field. *Phil Trans R. Soc. Lond. Ser. A.* **244**, 113–51.
42   Johnson, E. A., T. Murphy and O. W. Torreson 1948. Pre-history of the Earth's magnetic field. *Terr. Magn. Atmos. Elec.* **53**, 349–72.
43   Clegg, J. A., M. Almond and P. H. S. Stubbs 1954. The remanent magnetism of some sedimentary rocks in Britain. *Phil Mag. Ser. 7* **45**, 583–98.
44   Clegg, J. A., E. R. Deutsch and D. H. Griffiths 1956. Rock magnetism in India. *Phil Mag. Ser. 8* **1**, 419–31.
45   Blackett, P. M. S. 1956. Lectures on Rock Magnetism – Being the Second Weizmann Memorial Lecture, Dec. 1954, Jerusalem.
46   Kelvin, Lord (William Thomson) 1876. Presidential Address to the Section of Mathematics and Physics, in *Notices and abstract of misc. comm. to the Sections*, Brit Ass. Adv. Sci. Rep. 1–12.
47   Darwin, G. 1879. Precession of a viscous spheroid and the remote history of the Earth. *Phil Trans R. Soc. Lond.* **170A**, 447–538.
48   Barret, P. H. 1977. *Collected papers of Charles Darwin*, Univ. of Chicago Press, Chicago.
49   Brown, H. A. 1967. *Cataclysms of the Earth*. New York: Twayne.
50   Hapgood, C. H. 1958. *Earth's shifting crust – a key to some basic problems of Earth Science*. New York: Pantheon.
51   Gold, T. 1955. Instability of the Earth's axis of rotation. *Nature* **175**, 526–9.
52   Munk, W. H. 1956. Polar wandering: a marathon of errors. *Nature* **177**, 551–4.
53   Runcorn, K. 1955. Rock magnetism. *Adv. Phys.* **4**, 244.
54   Sullivan, W. 1974. *Continents in motion*. New York: Doubleday.
55   Irving, E. 1956. Palaeomagnetic and palaeoclimatological aspects of polar wandering. *Geofis. Pura. Applicata* **33**, 23–41.
56   Runcorn, S. K. 1959. Rock magnetism. *Science* **129**, 1002–11.
57   Cox, A. and R. Doell 1960. Review of palaeomagnetism. *Bull. Geol Soc. Am.* **71**, 645–768.
58   Cox, A. and R. Doell 1961. Palaeomagnetic evidence relevant to a change in the Earth's radius. *Nature* **189**, 45–7.
59   Carey, S. W. 1945. Tasmania's place in the geological structure of the world. Address to the Royal Society of Tasmania. 14 May.
60   Carey, S. W. 1976. *The Expanding Earth* (Developments in Geotectonics 10). Amsterdam: Elsevier.
61   Carey, S. W. (ed.) 1958. Continental drift: being a symposium on the present status of the continental drift hypothesis, held in the Geology Department of the University of Tasmania, in March 1956, Hobart.

Chapter 6

1   Vening-Meinesz, F. A. 1926. Gravity survey by submarine via Panama to Java, *Geogr. J.* **1xxi**, 144–59.
2   Vening-Meinesz, F. A. 1933. The mechanism of mountain formation in geosynclinal belts. *Proc. K. Acad. Wetensch.* vol. 36, 372–377.

References

3   Wiseman, J. D. H. and R. B. S. Sewell 1937. The floor of the Arabian Sea. *Geol Mag.* **74**, 219–30.

4   Darwin, F. 1887. Letter to Alexander Agassiz, 5 May. In *The Life and Letters of Charles Darwin*, J. Murray, London. vol. 2.

5   Bullard, E. C. 1975. The emergence of plate tectonics: a personal view. *Ann. Rev. Earth. Plan. Sci.* **3**, 1–30.

6   Wilson, J. T. 1983. Movements in earth science. *New Scientist* 26 Nov., 613–16.

7   Hess, H. 1962. Richard Montgomery Field. *Trans. Am. Geophys. Union* **43**, 1–3.

8   Schlee, S. 1973. *A history of Oceanography*. London: Robert Hale.

9   Wertenbaker, W. 1974. *The floor of the sea*. Boston: Little, Brown.

10 Hess, H. H. 1946. Drowned ancient islands of the Pacific Basin. *Am. J. Sci.* **244**, 772–91.

11 Hamilton, E. L. 1956. Sunken islands of the mid-Pacific mountains. *Geol Soc. Am. Mem.* **64**.

12 Ewing, M. and F. Press 1950. Crustal structure and surface wave dispersion. *Bull. Seis. Soc. Am.* **40**, 271–80.

13 Katz, S. and M. Ewing 1956. Seismic refraction measurements in the Atlantic Ocean. part VII: Atlantic Ocean Basin west of Bermuda. *Bull. Geol Soc. Am.* **67**, 475–510.

14 Bullard, E. C. 1954. A discussion on the floor of the Atlantic Ocean, Feb. 28th, 1953. *Proc. R. Soc. Lond. Ser. A.* **222**, 408–29.

15 Hess, H. H. 1955 The oceanic crust. *J. Mar. Res.* **14**, 423–39.

16 Ewing, M., J. Hirshman, and B. C. Heezen 1959. Magnetic anomalies of the mid-ocean ridge system. In *Preprints of Abstracts, 1st International Oceanographic Congress*, M. Sears (ed.), 24. Washington, DC; American Association for the Advancement of Science.

17 Worzel, J. L. and M. Ewing 1954. Gravity anomalies, and structure of the West Indies. *Bull. Geol Soc. Am.* **65**, 195–200.

18 Heezen, B. C., M. Tharp and W. M. Ewing 1959. *The floor of the oceans, 1: North Atlantic*. Geol Soc. Am. Spec. Paper No. 65.

19 Hess, H. H. 1962. History of ocean basins. In *Petrologic Studies: A Volume in Honor of A. F. Buddington*, A. E. J. Engel, H. L. James and B. F. Leonard (eds), 599–620. New York: Geology Society of America.

20 Dietz, R. S. 1961. Continent and ocean basin evolution by spreading of the sea-floor. *Nature* **190**, 854–7.

21 Ewing, M. and B. C. Heezen 1956. Oceanographic research programs of the Lamont Geological Observatory. *Geogr. Rev.* **46**, 508–35.

22 Menard, H. W. 1958. Development of median elevations in ocean basins. *Bull. Geol Soc. Am.* **69**, 1179–86.

23 Petersson, H. 1949. Exploring the bed of the ocean. *Nature* **164**, 468–70.

24 Revelle, R. and A. E. Maxwell 1952. Heatflow through the floor of the eastern North Pacific Ocean, with a comment by Sir Edward Bullard. *Nature* **170**, 199–200.

25 Bullard, E. C., A. E. Maxwell and R. Revelle 1956. Heat flow through the deep sea floor. *Adv. Geophys.* **3**, 153-81.

26 Mason, R. G. 1958. A magnetic survey off the west coast of the US between latitudes 32° and 36°N and longitudes 121° and 128°W. *Geophys. J. R. Astron. Soc.* **1**, 320–29.

27 Mason, R. G. and A. D. Raff 1961. A magnetic survey of the west coast of North America, 32°N – 42°N. *Bull. Geol Soc. Am.* **72**, 1259–65.

28 Menard, H. W. 1955. Deformation of the northeastern Pacific basin and the west coast of North America. *Bull. Geol Soc. Am.* **66**, 1149–98.

29 Vacquier, V. 1959. Measurements of horizontal displacements along faults in the ocean floor. *Nature* **183**, 452–3.

30 Girdler, R. 1958. The relationship of the Red Sea to the East African rift system. *Q. J. Geol. Soc. Lond.* **114**, 79–105.

31 Ewing, M., B. C. Heezen and J. Hirshman 1957. *Mid-Atlantic Ridge seismic belts and magnetic anomalies*. Seismol. Assoc. Intern. Union. Geodesy Geophys. Gen. Assembly Toronto (abstract).

32 Matanuyama, M. 1929. On the direction of magnetisation of basalt in Japan, Tyósen and Manchuria. *Proc. Imp. Acad. Jpn* **5**, 203–5.

References

33 Einarsson, Tr. 1957. Magneto-geological mapping in Iceland with the use of a compass. *Adv. Phys.* **6**, 232–9.

34 Khramov, A. N. 1957. Palaeomagnetism; the basis of a new method of correlation and subdivision of sedimentary strata. *Akad, Nauk. SSSR Doklady Geol.* **114**, 849–52 (English trans. (1958) 129–132).

35 Rutten, M. G. 1959. Palaeomagnetic reconnaissance of mid-Italian volcanoes. *Geol Mijn.* **21**, 373–4.

36 Cox, A., R. R. Doell and G. B. Dalrymple 1963. Geomagnetic polarity epochs and Pleistocene geochronometry. *Nature* **198**, 1049–51.

37 Girdler, R. W. and G. Peter 1960. An example of the importance of natural remanent magnetization in the interpretation of magnetic anomalies. *Geophys. Prospecting* **8**, 474–83.

38 Laughton, A. S., M. N. Hill and T. D. Allen 1960. Geophysical investigations on a seamount 150 miles north of Madeira. *Deep Sea Res.* **7**, 117–41.

39 Vine, F. J. and D. H. Matthews 1963. Magnetic anomalies over oceanic ridges. *Nature* **199**, 947–9.

40 Morley, L. W. and A. Larochelle 1964. Palaeomagnetism as a means of dating geological events. In *Geochronology in Canada* F. F. Osborne (ed.), 39–51. Roy. Soc. Canada Spec. Pub. No. 8. Toronto; University of Toronto Press.

41 Wilson, J. T. 1963. A possible origin of the Hawaiian islands. *Canadian J. Phys.* **41**, 863–70.

Chapter 7

1 Macelwane, J. B. 1950. *The Jesuit Seismological Association: 25 years.* St Louis, Missouri.

2 Bullen, K. E. 1958. Seismology in our atomic age (Presidential address at the International Association of Seismology and Physics of the Earth's Interior, Toronto, Sept, 1957.) *Comptes Rendus* **12**, 19. Assn. Seis. et Physique de l'Interieur de la Terre.

3 Hill, M. L. and T. W. Dibblee Jr 1954. San Andreas, Garlock and Big Pine Faults, California. *Bull. Geol Soc. Am.* **64**, 443–58.

4 Kennedy, W. Q. 1946. The Great Glen Fault. *Q. J. Geol Soc. Lond.* **102**, 41–76.

5 Wilson, J. T. 1952. Cabot Fault, an Appalachian equivalent of the San Andreas and Great Glen Faults and some implications for continental displacement. *Nature* **195**, 135–8.

6 Wilson, J. T. 1963. Hypothesis of the Earth's behaviour. *Nature* **198**, 925–9.

7 Wilson, J. T. 1963. Continental drift. *Sci. Am.* **208**, 86–100.

8 Bullard, E. C. 1964. Continental drift. *Q. J. Geol Soc. Lond.* **120**, 1–33.

9 Blackett, P. M. S., E. C. Bullard and S. K. Runcorn (eds.) 1964. A symposium on continental drift. *Phil Trans R. Soc. Lond. Ser. A.* **258**.

10 Jeffreys, Sir H. 1962. Comment on 'A suggested reconstruction of the land masses of the Earth as a complete crust'. *Nature* **195**, 448.

11 Sykes, L. R. 1963. Seismicity of the South Pacific Ocean. *J. Geophys. Res.* **68**, 5999.

12 Wilson, J. T. 1965. A new class of faults and their bearing on continental drift. *Nature* **207**, 343–7.

13 Vine, F. J. and J. T. Wilson 1965. Magnetic reversals over a young oceanic ridge off Vancouver Island. *Science* **150**, 485–9.

14 Dalrymple, G. B., R. R. Doell and A. Cox 1966. Recent developments in the geomagnetic polarity (epoch time scale). In *Abstracts for 1965*, 41. Geol. Soc. Amer. Spec. Paper No 87).

15 Heirtzler, J. R., X. Le Pichon and J. G. Baron 1966. Magnetic anomalies over the Reykjanes Ridge. *Deep Sea Res.* **13**, 427–43.

16 Heirtzler, J. R. and X. Le Pichon 1965. Crustal structure of the mid-oceanic ridges, pt (3): Magnetic anomalies over the Mid-Atlantic Ridge. *J. Geophys. Res.* **70**, 4013–33.

17 Pitman, W. C. III and J. R. Heirtzler 1966. Magnetic anomalies over the Pacific-Antarctic Ridge. *Science* **154**, 1164–71.

References

18 Vine, F. J. 1966. Spreading of the ocean floor – new evidence. *Science* **154**, 1405–15.

19 Glen, W. 1982. *The road to Jaramillo*, 339. Stanford University Press.

20 Bullard, E. C. 1975. The emergence of plate tectonics: a personal view. *Ann. Rev. Earth Plan. Sci. Lett.* **3**, 1–30.

21 Frankel, H. 1982. The development, reception, and acceptance of the Vine–Matthews–Morley hypothesis. *Historical Studies in the Physical Sciences* **13**, 1–39.

22 McKenzie, D. P. and R. L. Parker 1967. The North Pacific, an example of tectonics on a sphere. *Nature* **216**, 116–20.

23 Scheidegger, A. E. 1958. Seismological evidence for the tectonics of the Northwest Pacific Ocean. *Bull. Seism. Soc. Am.* **48**, 369–75.

24 Quennell, A. M. 1957. The structural and geomorphic evolution of the Dead Sea rift. *Q. J. Geol Soc., London* **114**, 1–18.

25 Morgan, W. J. 1968. Rises, trenches, great faults and plate motions. *J. Geophys. Res.* **73**, 1959–82.

26 Sykes, L. R. 1967. Mechanism of earthquakes and nature of faulting on the mid-oceanic ridges. J. Geophys. Res. **72**, 2131–53.

27 Wadati, K. 1934. On the activity of deep-focus earthquakes in the Japan Islands. *Geophys. Mag.* **8**, 305–25.

28 Benioff, H. 1954. Orogenesis and deep crustal structure: additional evidence from seismology. *Bull. Geol Soc. Am.* **65**, 385–400.

29 Menard, H.W . and R. S. Dietz 1951. Submarine geology of the Gulf of Alaska. *Bull. Geol Soc. Am.* **62**, 1263–85.

30 Raitt, R. W., R. L. Fisher and R. G. Mason 1955. Tonga Trench. *Geol Soc. Am. Spec. Paper* **62**, 237–254.

31 Isaacks, B., L. R. Sykes and J. Oliver 1967. The focal mechanisms of deep earthquakes in the Fiji–Tonga–Kermadec region (abstract). IUGG General Assembly, IASPEI, 177, Zurich.

32 Le Pichon, X. 1968. Sea-floor spreading and continental drift. *J. Geophys. Res.* **73**, 3661–97.

33 Isaacks, B., J. Oliver and L. R. Sykes 1968. Seismology and the new global tectonics. *J. Geophys. Res.* **73**, 5855–99.

34 McKenzie, D. P. 1969. Speculations on the consequences and causes of plate motions. *Geophys. J. R. Astron. Soc.* **18**, 1–32.

35 McKenzie, D. P. and W. J. Morgan 1969. Evolution of triple junctions. *Nature* **224**, 125–33.

36 Wilson, J. T. 1966. Did the Atlantic close and then re-open? *Nature* **211**, 676–81.

37 Dewey, J. F. and J. M. Bird 1970. Mountain belts and the new global tectonics. *J. Geophys. Res.* **75**, 2625–47.

38 Dewey, J. F. 1969. Evolution of the Appalachian/Caledonian orogen. *Nature* **241**, 124–9.

39 Karig, D. E. 1971. Origin and development of marginal basins in the Western Pacific. *J. Geophys. Res.* **76**, 2542–61.

40 Cann, J. R. 1974. A model for oceanic crustal structure development. *Geophys. J. R. Astron. Soc.* **39**, 169–87.

41 Morgan, W. J. 1971. Convection plumes in the lower mantle. *Nature* **230**, 42–3.

42 Dietz, R. 1964. Sudbury structure as an astrobleme. *J. Geol.* **72**, 412–34.

Chapter 8

1 Wilson, J. T. 1963. The movement of continents. Address presented at Symposium on the Upper Mantle Project, XIII General Assembly IUGG, Berkeley.

2 Wilson, J. T. 1967. A revolution in the Earth Sciences. Talk delivered in Ottawa, published in *Canadian Mining Metal. Bull.*

3 Roubault, M. 1953. Preface, *Sciences de la Terre.* **1**, i.

4 Maury, M. F. 1855. *The physical geography of the sea.* New York: Harper.

5 Sullivan, W. 1962. *Assault on the unknown.* London: Hodder & Stoughton.

6 Wilson, J. T. 1957. The crust. In *The planet Earth*, D. R. Bates (ed.) 48–73. London: Pergamon Press.

References

7   Wilson, J. T. 1961. *IGY: the year of the new moons.* Michael Joseph, London.
8   Greenberg, D. S. 1967. Mohole: the anatomy of a fiasco. In *The politics of pure science,* 170–206. New American Library, New York.
9   Hess, H. H. 1966. The Mohole Project, phase II. In *Drilling for Scientific Purposes,* Report of Symposium, Geol. Surv. of Canada, Paper 66–13.
10 Bascom, W. 1961. 'Core drilling under the ocean', letter from Jaggar to Field. Quoted in *A hole in the bottom of the Sea.* New York: Doubleday.
11 Wasserburg, J.
12 Schuchert, C. 1928. in *Theory of Continental Drift* van Waterschoot van der Gracht *et al.* (eds.) 104–44. Tulsa: American Association of Petroleum Geologists.
13 Holmes, A. 1953. The South Atlantic: land bridges or continental drift? *Nature* **171**, 669–71.
14 Sollas, W. J. 1903. The Figure of the Earth, *Q.J.G.S.* **59**, 180–8.
15 Mallet, R. 1858. Plate 12 in *Brit. Ass. Rep.* (original at Royal Society of London).
16 Mallet, R. 1873. Volcanic energy: an attempt to show its true origin and cosmical relations. *Phil Trans R. Soc. Lond.* **163**, 147–227.
17 Davison, C. 1927. *The founders of seismology.* Cambridge: Cambridge University Press.
18 Twain, M. 1872. *Mark Twain: Roughing It!*
19 Dutton, C. E. 1884. The Hawaiian islands and people. *Ordnance notes, No. 343,* Washington, April 23rd.
20 Duffield, W. A. 1972. A naturally occurring model of global plate tectonics. *J. Geophys. Res.* **77**, 2543–55.
21 Berman, A. 1983. The impact of oceanography on the military and security uses of the ocean in *Oceanography Present and Future,* P. G. Brewer (ed.) 205–16. Springer Verlag, Berlin.
22 Wertenbaker, W. 1974. *The floor of the sea.* Boston: Little, Brown.
23 Nierenberg, W. A. and R. Revelle 1980. *Edward Crisp Bullard (1907–1980).* Funeral address unpublished.

Chapter 9

1   Wright, W. B. 1923. The Wegener hypothesis. *Nature* **111**, 30–1.
2   Blackett, P. M. S., E. C. Bullard and S. K. Runcorn (eds.) 1965. A symposium on continental drift. *R. Soc. Lond. Phil Trans Ser. A* **258**, 1–323.
3   Davis, G. A., B. C. Burchfiel, J. E. Case and G. W. Viele 1974. A defense of an 'Old Global Tectonics'. In 'Plate tectonics, assessments & reconstructions', *Am. Assoc. Pet. Geol.* **23**. 16–23.
4   Baker, H. B. 1933. *The Atlantic rift and its meaning.* Detroit. Unpublished manuscript.
5   Marvin, U. B. 1973. Quote from Mather, 1971, in *Continental drift, the evolution of a concept,* 205. Smithsonian Institute.
6   Meyerhoff, A. A. and H. A. Meyerhoff 1972. The new global tectonics: major inconsistencies. *Bull. Am. Assoc. Pet. Geol.* **56**, 269–336.
7   Nitecki, M. H., J. L. Lemke, H. W. Pullman and M. E. Johnson 1978. Acceptance of plate tectonic theory by geologists. *Geology* **6**, 661–4.
8   Wilson, J. T. 1983. Movements in earth science. *New Scientist* 26 Nov., 613–16.
9   McPhee, J. 1983. In *Suspect Terrain,* Farrar, Strauss, Giroux New York.
10 Wilson, J. T. 1966. Abstract of 19th William Smith Lecture 'On a possible explanation for the crust's growth and movements'. *Proc. Geol Soc. Lond.* 42–3.
11 Dunham, K. C. 1967. Annual General Meeting. *Proc. Geol Soc. Lond.* 278–80.
12 Davison, C. 1927. *The founders of seismology.* Cambridge: Cambridge University Press.
13 Muir Wood, R. 1979. Is the Earth getting bigger? *New Scientist* 8 Feb, 287.
14 Wilson, J. T. 1960. Some consequences of expansion of the Earth. *Nature* **185**, 880–3.
15 Dutton, C. E. 1874. A criticism upon the contraction hypothesis. *Am. J. Sci.* **8** (3rd series), 113–23.
16 Tucker, A. 1983. Why On Earth. *Guardian,* 13 Jan.
17 Lyttleton, R. A. 1983. *The Earth and its mountains.* London: Wiley Interscience.

References

18 Kanaev, A. 1937. Speeches delivered at the opening of the 1937 XVII International Geological Congress, Moscow, A. A. Amrislanov, (ed.). Congress Reports, 158–74. Moscow.

19 Usov, M. A. 1937. *Phases and cycles of tectonogenesis of the west Siberian region.* In report of XVIII International Geological Congress, Moscow, Vol. 2, 623–8.

20 Obruchev, V. A. 1940. The oscillation hypothesis of geotectonics. *Izv. Akad. Nauk. SSSR. Ser. Geol. No. 1.*

21 Dudek, A. 1970. General Proceedings, Report of the XXIII IGC, Czechoslovakia, 172–89.

22 Kent, P. 1969. Prague – August 1968. *Proc. Geol Soc. Lond.* 219–24.

23 Beloussov, V. V. 1968. An open letter to J. Tuzo Wilson. *Geotimes* 13 (10), 17–19.

24 Beloussov, V. V. 1970. Against the hypothesis of sea-floor spreading. *Tectonophysics* 9, 489–511.

25 Beloussov, V. V. and Ye. Ye. Milanovskiy 1977. On tectonics and tectonic position of Iceland. *Tectonophysics* 37, 25–40.

26 Sullivan, W. 1974. *Continents in motion.* New York: Doubleday.

27 Kropotkin, P. N. 1967. The mechanism of crust movements (English version). *Geotectonics* 5, 276–85.

28 Kropotkin, P. N. 1971. Eurasia as composite continent. *Tectonophysics* 12, 261–6.

29 Belov, A. A. 1978. The tectonics of the Mediterranean Belt (English version). *Geotectonics* 12, 485–7.

30 Milanovsky, Ye. Ye. 1978. Some aspects of tectonic development and volcanism of the Earth in the Phanerozoic (pulsation and expansion of the Earth) (English version). *Geotectonics* 12, 403–11.

31 Khain, V. E. 1978. From plate tectonics to a more general theory of global tectonics (English version). *Geotectonics* 12, 163–76.

32 Beloussov, V. V. 1979. Why I do not accept plate tectonics. *Trans. Am. Geophys. Union (EOS)* 60, 207–11.

33 Kirillova, G. L. and L. I. Popeto 1980. Problems of tectonics at the 14th Pacific Science Congress (English version). *Geotectonics* 14, 328–332.

34 Kropotkin, P. N. 1981. Problems of dynamics of the lithosphere of the Earth and the planets in the light of the latest geophysical data (Proceedings of the 26th Int. Geol. Cong. Paris) (English version). *Geotectonics* 15, 190–3.

35 Rich, V. 1981. Open approval – Soviet plate tectonics. *Nature* 292, 489.

36 Demenitskaya, R. M. and A. M. Karasik 1969. The active rift system of the Arctic Ocean. *Tectonophysics* 8, 345–51.

37 JOIDES 1977. The future of scientific ocean drilling. Report of a Conference held at Woods Hole, Mass., 7–11 March 1977, Appendix 10, 91–2.

38 Il'in, A. V. 1978. Morphostructures of the ocean floor and some problems of the new global tectonics (English version). *Geotectonics* 12, 412–423.

Bibliography

Primary sources

Beloussov, V. V. 1962. *Basic problems in geotectonics*. New York: McGraw-Hill.
Bucher, W. H. 1933. *The deformation of the Earth's crust*. Princeton, NJ: Princeton University Press.
Burnet, T. 1681. *The sacred theory of the Earth*. London: W. Kettilby.
Daly, R. A. 1914. *Igneous rocks and their origins*. New York and London: McGraw-Hill.
Darwin, C. R. 1842. *The structure and distribution of coral reefs*. London: Smith, Elder and Co.
Darwin, C. R. 1859. *On the origin of species . . . .* London: John Murray.
Du Toit, A. L. 1937. *Our wandering continents: an hypothesis of continental drift*. Edinburgh and London: Oliver and Boyd.
Fisher, O. 1881. *The physics of the Earth's crust*. London: Macmillan.
Frankel, H. 1982. The development, reception, and acceptance of the Vine-Matthews-Morley hypothesis. *Hist. Stud. Phys. Sci.* **13**, 1–39.
Gass, I. G., P. J. Smith, R. C. L. Wilson (eds) 1971. *Understanding the Earth*. Sussex: Artemis Press.
Green, W. L. 1875. *Vestiges of the molten globe*. London: E. Stanford.
Green, W. L. 1887. *Vestiges of the molten globe: volume 2, the volcanic problem*. Honolulu, Hawaii: privately published.
Hallam, A. 1973. *A revolution in the Earth sciences*. Oxford: Clarendon Press.
Holmes, A. 1966. *Principles of physical geology, 2nd edn*. London: Thomas Nelson.
Jeffreys, H. 1924. *The Earth*. Cambridge: Cambridge University Press.
Joly, J. 1909. *Radioactivity and geology*. London: Constable.
Joly, J. 1925. *The surface history of the Earth*. Oxford: Clarendon Press.
Kober, L. 1921. *Der Bau der Erde*. Berlin: Gebrüder Borntraeger.
Kuhn, T. 1962. *The structure of scientific revolutions*. Chicago: University of Chicago Press.
Le Pichon, X., J. Francheteau and J. Bonnin 1973. *Plate tectonics*. Amsterdam: Elsevier.
Lyell, C. 1830–3. *Principles of geology, 3 vols*. London: John Murray.
Lyell, C. 1838. *The elements of geology*. London: John Murray.
Mather, K. F. 1967. *A source book in geology 1900–1950*. Harvard, Mass.: Harvard University Press.
Pepper, J. H. 1861. *The playbook of metals*. London and New York: Routledge, Warne and Routledge.
Routh, E. J. 1884. *The advanced part of a treatise on the dynamic system of rigid bodies, 4th edn*. London and New York: Macmillan.
Runcorn, S. K. (ed.) 1962. *Continental drift*. London: Academic Press.
Snider–Pellegrini, A. 1857. *La Création et ses mysteres dévoilés*. Paris: Franck and Dentu.
Suess, E. 1875. *Die enstehung der Alpen*. Vienna: W. Braumüller.
Suess, E. 1885–1909. *Das antlitz der Erde, 4 vols*. Vienna: Prag F. Tempsky.
Sverdrup, H. U., M. W. Johnson and R. H. Fleming 1946. *The Oceans*. New York: Prentice Hall.
Velikovsky, I. 1952. *Worlds in collision*. New York: Doubleday, Garden City.
Velikovsky, I. 1955. *Earth in upheaval*. New York: Doubleday, Garden City.

Verhoogen, J., F. J. Turner, L. E. Weiss, C. Wahraftig and W. S. Fyfe 1970. *The Earth*. New York: Holt, Rinehart and Winston.

Wallace, A. R. 1876. *The geographical distribution of animals*. London: Macmillan.

Wegener, A. L. 1915. *Die enstehung der kontinente und ozeane*. Braunschweig: P. Vieweg.

Wegener, A. L. 1924. *The origin of the continents and oceans* (translation of the 3rd edn). London: Methuen.

Secondary sources

Bates, C. C., T. F. Gaskell and R. B. Rice 1982. *Geophysics in the affairs of man*. Oxford: Pergamon Press.

Bolt, B. A. 1976. *Nuclear explosions and earthquakes*. San Francisco: W. H. Freeman.

Burchfield, J. D. 1975. *Lord Kelvin and the age of the Earth*. London: Macmillan.

Gillispie, C. C. 1951. *Genesis and Geology*. New York: Harper and Row.

Glen, W. 1982. *The road to Jaramillo*. Stanford, California: Stanford University Press.

Greene, M. T. 1982. *Geology in the Nineteenth Century: Changing views of a changing world*. Ithaca: Cornell University Press.

Marvin, U. B. 1973. *Continental drift: the evolution of a concept*. Washington: Smithsonian Institution Press.

Porter, R. 1977. *The making of geology*. Cambridge: Cambridge University Press.

Pyne, S. J. 1980. *Grove Karl Gilbert*. Austin: University of Texas Press.

Rupke, N. A. 1970. Continental drift before 1900. *Nature* **227**, 349–50.

Schlee, S. 1973. *A History of Oceanography*. London: Robert Hale.

Sengör, A. M. C. 1982a. Eduard Suess' relations to the pre-1950 schools of thought in global tectonics. *Geol, Rundschau* **71**, 381–420.

Sengör, A. M. C. 1982b. Classical theories of orogenesis. In Miyashiro, A., K. Aki and A. M. C. Sengör *Orogeny*. Chichester: John Wiley.

Sullivan, W. 1974. *Continents in motion – the New Earth Debate*. New York: McGraw Hill.

Reference works

Dictionary of Scientific Biography, 16 vols 1970–8. Ed. C. C. Gillispie, New York: Charles Scribner's Sons.

Great Soviet Encyclopedia, transl. 3rd edn, 26 vols. 1973. London: Macmillan.

Sarjeant, W. A. S. 1980. *Geologists and the history of Geology*, 4 vols. London: Macmillan.

Index